AF704484

Promoting health and productivity in the computerized office:

Models of successful ergonomic interventions

Promoting health and productivity in the computerized office:

Models of successful ergonomic interventions

Edited by

Steven L. Sauter

National Institute for Occupational Safety and Health

Marvin J. Dainoff

Miami University

and

Michael J. Smith

University of Wisconsin

Taylor & Francis

London • New York • Philadelphia

UK Taylor & Francis Ltd. 4 John St. London WC1N 2ET

USA Taylor & Francis Inc., 1900 Frost Road, Suite 101, Bristol, PA 19007

Copyright © Taylor & Francis Ltd., 1990

All rights reserved. No part of this publication may be reproduced, stored in a retrieval system, or transmitted, in any form or by any means, electronic, electrostatic, magnetic tape, mechanical, photocopying, recording or otherwise, without the prior permission of the copyright owners.

British Library Cataloguing in Publication Data
Promoting health and productivity in the computerized office.
1. Computer systems. Health aspects
I. Sauter, Steven L. II. Dainoff, Marvin J. (Marvin Julian) III. Smith, Michael J.
613.62

ISBN 0-85066-756-9

Library of Congress Cataloging-in-Publication Data
Promoting health and productivity in the computerized office: models of successful ergonomic interventions/editors, Steven L. Sauter, Marvin J. Dainoff, Michael J. Smith.
p. cm.
Includes bibliographical references and index.
ISBN 0–85066–756–9
1. Video display terminals—Health aspects. I. Sauter, Steven L.
II. Dainoff, Marvin J. III. Smith, Michael J. (Michael James),
RC965.V53P77 1990
616.9′ 803—dc20 90–44336
CIP

Phototypesetting by
Chapterhouse, The Cloisters, Formby, England.
Printed in Great Britain by Burgess Science Press.

Contents

Preface

A 1977 report (Moss, 1977) from the National Institute for Occupational Safety and Health (NIOSH) was probably the first acknowledgment in the USA of ergonomic stressors in video display terminal (VDT) work. Since this observation, an enormous literature and four major international conferences have focused on health and ergonomic aspects of VDT use. NIOSH, alone, has conducted three dozen health hazard evaluations and published six dozen scientific reports on this subject (see NIOSH, 1987, for a bibliography and summary of this work). Nevertheless even casual observation reveals that freedom from ergonomic stressors has yet to become the norm in workplaces where VDTs are used.

Abatement of occupational or environmental health hazards is defined, classically, in terms of the following sequences of events: (1) problem identification and evaluation, and (2) development, dissemination and implementation of control technology. In terms of resolving health problems associated with ergonomic aspects of VDT use, progress is most clearly evident in the first stages of this process, i.e. the ergonomic determinants of somatic disturbances such as eye and musculoskeletal disturbances are generally understood.[1] Although significant strides are also evident in the development, dissemination, and implementation of control technology (e.g. see American National Standard for Human Factors Engineering of Visual Display Terminal Workstations, 1988), we believe further progress in this area has been impeded by the paucity of information providing (1) direction for management of ergonomic reforms in VDT work, and (2) incentives for undertaking ergonomic reform.

Control information presently available consists primarily of standards or rules for the design of VDT workstations and workplaces. Ergonomic reform in the VDT workplace, however, is more than the perfunctory implementation of rules; it is a costly and complex process requiring careful management. Unfortunately, the literature is nearly devoid of demonstration studies or case reports addressing management policy and procedures underlying successful redesign of VDT work and work-

[1] Important questions still remain regarding the etiology of chronic pain syndromes among VDT users.

places. In our view, what is needed is increased analysis, delineation and dissemination of management models leading to successful ergonomic reforms to provide *principled* guidance for further intervention. At the same time, systematic analysis of the benefits of ergonomic interventions in terms of worker well-being or other bottom-line, organizational performance measures is needed to motivate and justify or validate such reforms.

In order to expand this frontier of information, a NIOSH-sponsored workshop 'Promoting Health and Productivity in the Computerized Office: Models of Successful Ergonomic Interventions' was organized early in 1988, and hosted by the Center for Ergonomic Research, Miami University, Oxford, Ohio, in August of that year. The aim was to identify and critically examine applied research on the design and pay-off of ergonomic interventions in VDT workplaces.

The solicitation for contributions to the workshop resulted in an array of intervention reports from all points of the globe. Topics addressed by these reports included the following:

(a) effects of physical ergonomic and job/task redesign interventions in the VDT workplace,
(b) management approaches (policies and procedures) underlying successful interventions in VDT workplaces, and
(c) clinical techniques for assessing and managing health disturbances among VDT users.

The papers presented in this volume are largely, but not exclusively, the products of this workshop. They are organized into three sections corresponding to the topical areas identified above.

The empirical research reported here ranges from small, controlled laboratory experiments to broad, prospective, demonstration studies in field situations. A variety of interventions, including physical modification of workstations and work environments, reorganization of work tasks, and training and modification of work habits are represented. Some of this research is not immune to criticism for shortcomings in scientific rigour. The research designs are sometimes only quasi-experimental at best, and in several cases attribution of effects is complicated by concurrent changes in both work organization and the physical environment, or by unanticipated events which intervened in the course of the study. These constraints represent seemingly inevitable consequences of prospective research in real work settings. Nevertheless, from a meta-analytic perspective, the implications of this body of work are compelling. Collectively, the studies reported here show a matrix of effects suggesting significant health and productivity gains from an intelligent application of ergonomics in the VDT workplace.

Several of the studies (e.g. Dainoff; Francis and Dressell; Ong; Thompson) converge, for example, in demonstrating rather striking productivity increases (and rapid recoupment of investment costs) in the range of 20% resulting from ergonomic improvements in the design of VDT jobs and workplaces. Dramatically greater increases are described in Sullivan's review of intervention research at Aetna Life and Casualty. These papers also report concomitant increases in workplace satisfaction (e.g. Francis and Dressell; Sullivan) and decreases in somatic complaints (e.g. Dainoff; Ong).[2] Additionally, marked reductions in absenteeism, labour turnover, sick leave, and compensation claims are described (e.g. Aarås *et al.*; Ong; Sullivan; Thompson)

Underlying successful corporate responses to the stresses posed by VDT/computer technology are management philosophies and practices which appear rare in the contemporary business-world. Östberg and Högberg describe a 'Swedish model', characterized by labour-management codetermination of working conditions, which has proven successful at the Swedish Telecommunications Administration and, apparently, is the norm for Sweden. While a formal, social democratic approach of this type is unusual in the USA, the report by Westin describes management practices at Federal Express which are similar in some respects to the Swedish model, and which appear to have effectively eliminated VDT work as a job stress factor in that company. Specifically, both situations are characterized by a 'macroergonomic' management approach which recognizes the implications of VDT technology for the entire system of work (e.g. equipment/facilities design, work organization, health and safety, compensation, etc.), and the need for a coordinated management response in all sectors. Other core ingredients included top corporate commitment to ergonomics, high levels of labour-management communication and cooperation, and ergonomics training throughout the organization. (The importance of these factors is revealed also in the reports by Baitsch and by Lindström.) Additionally, follow-up evaluation or field trials of work modifications were important management practices at both Federal Express and the Swedish Telecommunications Administration. The report by Berns and Claridge provides a rather striking demonstration of the general need for improvement in these evaluation practices.

Finally, unlike the focus in the majority of interventions described in this volume, the reports by Patkin and by Sheedy and Parsons remind us the potential role of predisposing individual factors or work habits (e.g. pre-existing visual deficits or excessive muscle tension) in the etiology of disorders associated with ergonomic aspects of the workplace. Rather than absolving management from responsibility in such cases, the two papers show how working conditions and management practices can play an important role in both the potentiation and the rehabilitation of these cases.

Our commentary here represents only a snapshot of the broad treatment of ergonomic interventions in VDT work contained in this volume. The present reports provide both a strong rationale and general framework for such interventions. We hope they will serve to stimulate and facilitate more widespread ergonomic reform in the VDT workplace.

Steven L. Sauter
Marvin J. Dainoff
Michael J. Smith

[2] Productivity and comfort effects are less pronounced in the chapter by Winkel and Oxenburgh, owing to the nature of the intervention (intermittent physical activity which distracted from worktime) in the presence of an already advanced ergonomic environment.

References

American National Standards Institute, 1988, *American National Standard for Human Factors Engineering of Visual Display Terminal Workstations ANSI/HFS Standard No. 100–1988*, (Santa Monica, CA: Human Factors Society).

Moss, C. E., Murray, W. E., Parr, W. H., Messite, J. and Karches, G., 1977, *A Report on Electromagnetic Radiation Surveys of Video Display Terminals*. Report No. 78–129 (Cincinnati, OH: National Institute for Occupational Safety and Health) NTIS No. PB–297–823.

NIOSH, 1987, Publications on Video Display Terminals (Cincinnati, OH: National Institute for Occupational Safety and Health).

Contributors

Arne Aarås
Alcatel STK A/S
Postboks 60 Okern
0508 Oslo 5
Norway

Christof Baitsch
Work and Organizational Psychology Unit
Swiss Federal Institute of Technology
ETH — Zentrum CH-8092
Zurich
Switzerland

Thomas Berns
Ergolab-Ergonomilavoratoriet
Eriksbergsgatan 10B
S-114 30 Stockholm
Sweden

Nigel Claridge
NOMOS Management AB
Box 2013
S-182 02
Danderyd
Sweden

Marvin J. Dainoff
Center for Ergonomic Research and Department of Psychology
Miami University
Oxford
OH 45056
USA

David L. Dressel
Office of Facility Planning and Management
University of Illinois
807 S Wright Street
Champaign
IL 61820
USA

JoEllen Francis
Office of Facility Planning and Management
University of Illinois
807 S Wright Street
Champaign
IL 61820
USA

Yngve KO Högberg
Marketing Department
Swedish Telecommunications Administration
123 86 Farsta
Sweden

Stig Larsen
Medstat
Stromsveien 49–53
Postboks 37
N-2011 Strommen
Norway

Karl Lindström
Department of Psychology
Institute of Occupational Health
Topeliuksenkatu 41a A
SF-00250 Helsinki
Finland

Choon-Nam Ong
Department of Community, Occupational and Family Medicine
National University of Singapore
Lower Kent Ridge Road
Singapore 0511

Olov N. Östberg
Swedish Agency for Administrative Development
PO Box 34107
S-10026 Stockholm
Sweden

Maurice Oxenburgh
Occupational Health and Safety
77 Ryan Street
Lilyfield
NSW 2040
Australia

Michael Patkin
PO Box 735
Whyalla
SA 5600
Australia

Steven Parsons
School of Optometry
University of California, Berkeley
Berkeley
CA 94720
USA

James E. Sheedy
School of Optometry
University of California, Berkeley
Berkeley
CA 94720
USA

Elinar Stranden
Head of the Vascular Laboratory
Aker Hospital
0514 Oslo 5
Norway

Carol Sullivan
Corporate Planning and Effectiveness
Aetna Life and Casualty
151 Farmington Avenue
Hartford
CT 06156
USA

David A. Thompson
Department of Industrial Engineering and Engineering Management
Terman Engineering Center
Stanford University
Stanford
CA 94305
USA

Rolf Westgaard
University of Trondheim
Norwegian Institute of Technology
Division of Organization and Work Science
Alfred Getz vei 1
N 7034 Trondheim
Norway

Alan F. Westin
Department of Political Science
Columbia University
New York
NY 10027
USA

Jorgen Winkel
National Institute of Occupational Health
Department of Physiology
Division of Applied Work Physiology
S-171 84 Solna
Sweden

Disclaimer

The contents of this volume are reproduced herein as received from contributors. The opinions, findings and conclusions expressed are not necessarily those of the National Institute for Occupational Safety and Health (NIOSH), nor does mention of company or product names constitute endorsement by NIOSH.

Acknowledgements

We are grateful to Rani Lueder (Humanics), Dick Shell (University of Cincinnati), Alex Cohen, Larry Fine, Tom Hales, and Shiro Tanaka of NIOSH, and to the many other reviewers for their suggestions and encouragement in the preparation of this book.

PART I
Physical ergonomic interventions and job/task redesign

Chapter 1

Workspace Influence on Worker Performance and Satisfaction: An Experimental Field Study

JOELLEN FRANCIS AND DAVID L. DRESSEL

Introduction

Research evidence indicates that planning a comfortable work environment which facilitates employee and organizational effectiveness is a good investment. According to Wineman (1986), personnel salaries and benefits are estimated to comprise 90–92% of the overhead costs for an office building over a 40-year life span. The remaining 8–10% of the costs include maintenance, utilities, and start-up building and equipment costs. Based upon extensive research on productivity and office design, Brill *et al.* (1984) reported the ratio of personnel costs to building costs to be 13 : 1.

A relatively small investment in planning and implementing improvements in the workplace has been shown to yield a substantial return where subsequent performance improvements were meaningfully quantified. The American Productivity Center, sponsored by Steelcase, Inc., conducted a survey in 1982 of 99 US companies which implemented efforts to improve employee productivity (American Productivity Center, 1982). The respondents reported using performance measures including work output, error and task difficulty reduction, product and service quality, response time, office space utilization, and staffing requirements. Where environmental design changes were made, the companies reported paybacks for their efforts ranging from 6 months to 24 months.

Environmental variables which have been reported to influence employee performance include the following: lighting, temperature, noise, workstation, degree of workstation enclosure, and workspace layout (Brill *et al.*, 1984; Springer, 1986; Sundstrom, 1986). Several studies which implemented and evaluated systematic efforts to improve workstation furniture have reported significant improvements in worker performance and subjective ratings of satisfaction with the workplace. In a laboratory study of 118 visual display terminal (VDT) operators at an insurance company, Springer (1986) measured data entry and dialogue tasks at both traditional and ergonomically designed workstations. Existing conditions in the work environment and actual VDT tasks were simulated in four experimental conditions in which only furniture was varied, and in a control condition which used existing furniture. A 15% improvement in data entry (10% in the dialogue task) was found in comparing the best case ergonomic furniture condition to the traditional furniture. Workstations which permitted the user to adjust the VDT display, keyboard and task chair were preferred by the employees. Comfort, adjustable furniture, VDT screen surface and keyboard surface height were workstation factors rated to be of greatest importance to the employees. A payback for the difference in cost between a standard and adjustable workstation was calculated considering performance improvement, depreciation and after-tax cash flow (22 weeks for data entry; 36 weeks for dialogue).

In a similar laboratory study of VDT tasks, Dainoff *et al.* (1982; this volume) found a statistically significant difference ($P<0\cdot01$) in worker performance between two conditions in which the workstation furniture was manipulated to be ergonomically correct and incorrect. Thirteen subjects performed simulated data entry tasks for 3-h sessions over 2 days in each experimental condition. Height of the chair, keyboard and monitor and illumination in two rooms were individually adjusted for each subject to produce maximally and minimally comfortable work environments.

A 17-month field study sponsored by Westinghouse Furniture Systems Division and conducted by Human Factors/Industrial Design, Inc., compared performance and workplace satisfaction of 122 insurance clerical workers before and after installation of new open office systems furniture (Westinghouse Architectural Systems Division, 1982). The office renovation was planned by the insurance group's architectural design and facility management team. A control group was comprised of similar workers who did not move from their existing bull pen arrangement of traditional furniture. The group receiving new furniture and a new layout demonstrated a 5·5% increase in productivity ($P<0\cdot025$) and satisfaction with the workspace ($P<0\cdot01$) following the move to new offices. The new workstations were partitioned and provided with task lighting. Workstation components were selected and a layout developed according to employee task requirements. Control group performance and satisfaction did not significantly change over the 17-month period under study. A simple payback on the furniture was calculated to be 23 months. Employee satisfaction significantly improved from before to after furniture installation with 26 of 30 workplace features.

The present study analysed worker performance and office satisfaction before and after new task-oriented workstations were installed in two work groups. The study was conducted by the authors while at the US Army Construction Engineering Research Laboratory (USA CERL) (Francis *et al.*, 1986; Dressel and Francis, 1987). A US government agency commissioned the investigation to determine the costs and benefits of improving its office facilities. Workspace requirements of the agency's procurement unit were measured by a task-analysis questionnaire and interviews. A design of workstations and work group layout was developed for two typical work groups, each comprising 17 procurement agents and support staff. The type of furniture was varied between two similar work groups: one treatment group (E1) received new, ergonomically-designed systems furniture and the other treatment group (E2) was given a new arrangement and additional items of their existing conventional metal furniture. Performance and satisfaction data were compared between experimental and control groups before and after furniture installation. It was hypothesized that the two treatment groups would display a significant increase in performance and office satisfaction compared with the control group, as a result of improved workstation design and layout. Savings realized from increased performance were expected to exceed the cost of the systems furniture over the 21-month study period.

Method

Subjects

The subjects of study were procurement agents and trainees within the Contracting and Production Division of a government agency whose function was to procure construction supplies. Job-related activities were similar within and between separate work groups within the Division, distinguished only by the type of materials procured. Each work group within the Division comprised a supervisory procurement agent, eight to thirteen procurement agents, one to three trainees, one secretary, and no or one procurement clerk.

Assignment of Division work groups to treatment and control groups was random, with the provision that only fully staffed and equivalently performing groups were eligible for the study. An analysis of variance of historical performance scores for all work sections within the Division was conducted and groups were contrasted a posteriori by means of a Duncan's multiple range test in an effort to select the most comparable groups. Following these analyses two work groups (E1 and E2) were selected for the experimental manipulation. Each of these two groups comprised 14 procurement agents and trainees. A control group from the same Division consisting of four work sections (C1, C2, C3, C4) was also selected. The number of procurement agents and trainees in each section ranged from nine to sixteen. The total number of procurement agents and trainees in the four control sections was 46; the experimental groups' subjects totalled 28. Employee salary was distributed similarly between all work groups. The average employee age was 44 years and average length of employment was 12·5 years. Approximately 67% of the subjects were female. The total number of subjects in all groups combined was 74.

Two of the four work sections (C1 and C2) constituting the performance data control group also served as control groups for the questionnaire data.

Procedure

Arrangements were made with the agency management for the study to be conducted in an unobtrusive manner. The Assistant Division Chief agreed to provide computerized monthly performance records and to administer the office satisfaction questionnaire at the designated times. Only performance records for procurement agents and trainees were analysed as it was judged by the researchers that performance of supervisory and support personnel was more difficult to measure and existing records may not have been meaningful.

Subjects were aware that USA CERL was conducting a study of their agency's offices, but were not cognizant that performance measures would be analysed as part of the study. Upper management agreed to withhold any work and/or policy changes for this Division of employees until data collection was completed.

Information regarding agency operations, policies, communications and facility requirements was obtained by administering questionnaires to all Division employees and interviewing work group supervisors and the assistant Division head. Office requirements data elicited from the employees included identification of job activities, usage of materials and equipment within and outside of the workstation, amount of time spent in the workstation, and privacy and co-worker accessibility needs. Interview data provided detailed information on organizational and work group functions, work flow and communications patterns, management policies and goals, and requirements for workstation location relative to the location of co-workers' workstations.

Design and layout of workstations were developed on the basis of requirements information collected. Two work groups of 17 employees each (14 agents/trainees, 1 supervisor, 1 secretary and 1 clerk) were selected to receive redesigned workspaces. The type of furniture was varied between the two groups while the workstation design and layout were matched as closely as possible between the groups. E1 received new systems furniture (panel-hung components) and E2 was given a new arrangement and additional components of the existing standard metal furniture. Furniture was installed during a weekend to minimize work disruption. Furniture installation occurred in August 1983. Conditions were unchanged for the control group.

Workstation design and layout

The existing workstations were not partitioned and typically consisted of a standard double pedestal metal desk, desk chair and a five-drawer file cabinet (Figure 1.1). Desks had been painted orange or gold and laminated wood surfaces were placed on top of the desks as part of an earlier attempt by the agency to improve the furniture's appearance. The arrangement of desks varied between groups and often appeared to be haphazardly placed. Centrally located computer terminals were shared, but were not located in individual workspaces.

With the assistance of consulting architects workstations were designed and furniture selected to provide an optimum working environment which satisfied the occupants' requirements within space and budget constraints. The new workstations contained a greater amount of storage and reference document space which was more easily accessible than the existing situation (Figures 1.2 and 1.3). Partitions were added to permit greater acoustical and visual privacy and to define the work group. A second, rear, worksurface was provided in anticipation of the planned addition of computers which were to be shared between every two procurement agents. Task lighting mounted under the overhead document cabinets (systems furniture) and under desktop document organizers (conventional furniture) illuminated the worksurfaces. A small partitioned conference area and side chairs shared between every two procurement agents allowed for single visitors and small group discussions.

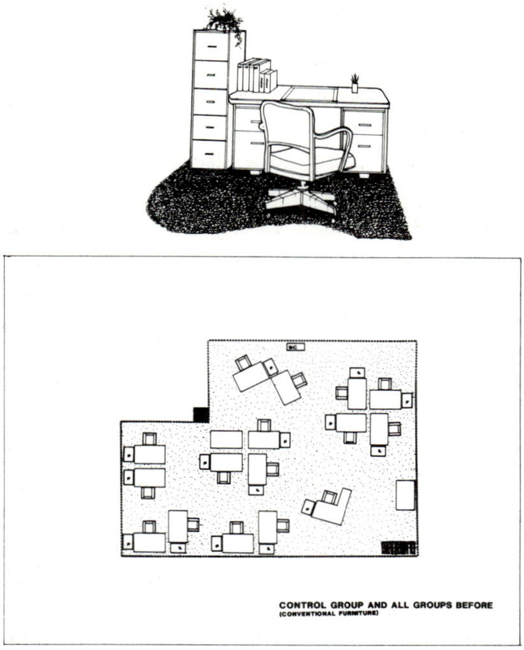

Figure 1.1. Typical existing furniture and workstation layout.

Only the systems furniture group (E1) received new ergonomic workstation chairs with back and height adjustments. The new systems furniture was tan with brown accents.

In order to provide a similar amount of worksurface and storage space between the two experimental groups, the conventional furniture workstation required a larger footprint than the panel-hung systems furniture workstation. The redesigned workstation for the procurement agent was 50 ft^2 in the improved conventional work group and 40 ft^2 in the systems work group.

All groups were housed in the same office building, although not immediately adjacent to one another. All groups were aware of the design changes made in the two treatment groups but not that performance measurements between groups were being compared.

Performance data

The unit of measurement selected was the number of purchase request line items completed per month. The agency regularly maintained records of the type and amount of procurement actions and workhours expended for each type of action for each work group. Small purchase request line items were recorded separately from large purchase request line items, since more effort was required to procure large

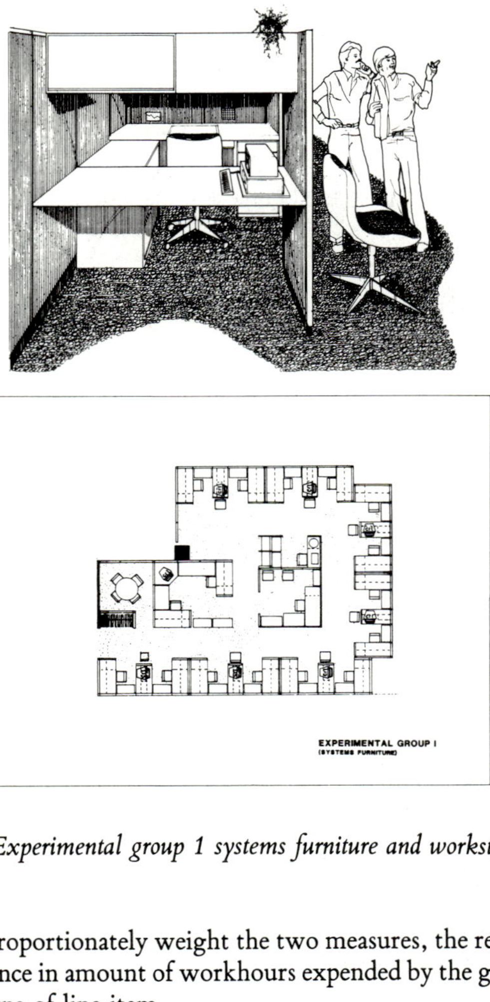

Figure 1.2. Experimental group 1 systems furniture and workstation layout.

items. In order to proportionately weight the two measures, the researchers analysed the historical difference in amount of workhours expended by the groups under study to complete each type of line item.

A composite performance measure was developed, the combined productivity index (CPI), which equated large and small item efforts using a coefficient of 2·89. The formula is as follows:

$$\text{CPI} = \frac{\text{number of small items} + (2\cdot 89)\ \text{number of large items}}{\text{workhours}}$$

Workhours consisted of the actual number of working hours expended, excluding sick leave, vacation leave or other compensated absences. Total sick leave hours were also recorded for the groups under study.

A measure of backlog was documented for each group which represented the number of purchase request line items which had not been completed by the end of the month.

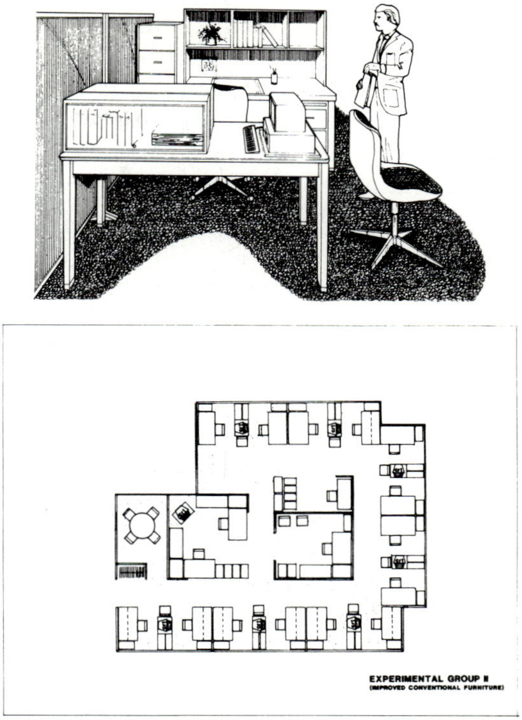

Figure 1.3. Experimental group 2 improved conventional furniture and workstation layout.

Performance data were gathered monthly and analysed over a period of 21 months from October 1982 to June 1984. Data were recorded 10 months prior to the experimental manipulation to form a baseline against which performance over 11 months following furniture installation was compared for each group.

Office satisfaction

An office satisfaction questionnaire was administered on three different occasions: 6 months before, 1 month after and 9 months after furniture installation. Respondents 6 months before the furniture installation included the two experimental groups and two control groups (C1 and C2). Only one of the control groups (C1) was included in the second questionnaire administration (1 month after) to control for the effect of two 'post' administrations. The questionnaire was administered on a third occasion to all four groups (9 months after) to control for the effects of the novelty of the workspace.

Respondents were asked to rate along a five-response scale (highly satisfied to highly dissatisfied) their satisfaction with features of their workstation and the office building in which they worked. Workstation features listed included furniture, control over visual and noise distractions, ease of communication with and access to co-workers, amount of floor area, worksurfaces, chair and filing/storage space. Facility features included lighting, conference facilities, circulation of people, heating and air conditioning, air circulation and electrical outlets.

Following each of the lists of features was an item asking respondents to rate their general level of satisfaction with their workstation and their office building. A total of 37 features were listed on the questionnaire. Respondents were also asked to rank order the five features they considered to be most important from the list.

Results

Performance

Analysis of performance changes (Student's *t*) from E1 (systems furniture) supported the hypothesis that improving the workspace would result in significantly increased performance ($p<0\cdot025$). This group displayed a 20·6% increase in performance level as compared against its baseline performance before furniture installation. E2 (improved conventional furniture) performance did not significantly increase from before to after installation of the furniture (4% increase). Also, the control groups' performance levels did not significantly change from before to after (0·6% increase). Figures 1.4 and 1.5 display the changes in performance.

Sick leave data did not significantly change for any of the groups over the period of the study.

Backlog data (purchase requests not completely processed by the end of each month) were analysed to test the extent of variation in the amount of work to be done each month and its relationship with actual performance. Backlog data showed a significant positive correlation with the productivity index over all groups and months prior to furniture installation ($r=0\cdot37$; $P<0.001$) but not for the 11 months following furniture installation ($r=0\cdot06$). Correlation of backlog with productivity was positive for E1 before ($r=0\cdot82$; $P<0\cdot01$) and negative (but not statistically significant) after furniture installation. The correlation was reversed for E2 (before: non-significant, negative; after: $r=0\cdot68$; $P<0\cdot01$). The only significant result for the control groups was a negative correlation ($r=-0\cdot40$; $P<0\cdot05$) over all months in the study. Backlog showed a significant increase from before to after installation for E1 and the control groups ($P<0\cdot05$), but not for E2. At no time was the backlog for any group equal to or near zero. The backlog analyses did not indicate a clear relationship between productivity and backlog, either in magnitude or direction.

Statistical testing for quarterly trends in the performance data produced no significant results.

Office satisfaction

The satisfaction questionnaire results generally confirmed the hypothesis that improving workplace design increases ratings of satisfaction with the workplace.

Pre- and post-installation comparisons of mean responses (Student's *t*) to the question of the general level of satisfaction with the workstation for E1 showed

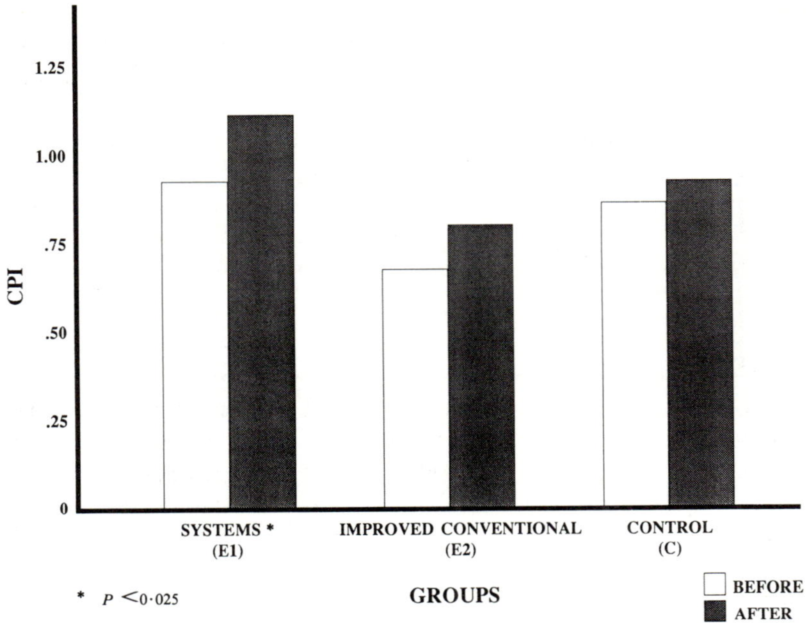

Figure 1.4. Combined productivity index before and after furniture installation.

significant increases both 1 month and 9 months following furniture installation ($p<0 \cdot 001$). E2 also displayed a similar significant improvement in workstation satisfaction ($p<0 \cdot 01$). However, only E1 also showed significantly increased satisfaction, in general, with the facility ($p<0 \cdot 001$). The average response given by E2 for the facility satisfaction item was 'neutral' on all questionnaire administrations. Ratings of general satisfaction by respondents in both control groups did not significantly differ across all questionnaire administrations.

Figure 1.6 displays the mean levels of general satisfaction with the workstation and the facility before and after workstation installation for each group.

The items most frequently rated as important to respondents from all groups combined included: control over noise, lighting (brightness level and task lighting), air circulation, heating and air conditioning, workstation chair, and control over visual distractions. E1 was significantly more satisfied with all of these features except heating and air conditioning following furniture installation ($p<0 \cdot 05$). Of these features, E2 gave significantly improved post-installation satisfaction ratings to 'control over noise', 'task lighting', and 'control over visual distractions' ($p<0 \cdot 025$). Although not statistically significant, it is noteworthy that the mean satisfaction ratings by one or both of the control groups decreased for those features which were manipulated in the experimental groups (noise, lighting, chair and visual distractions).

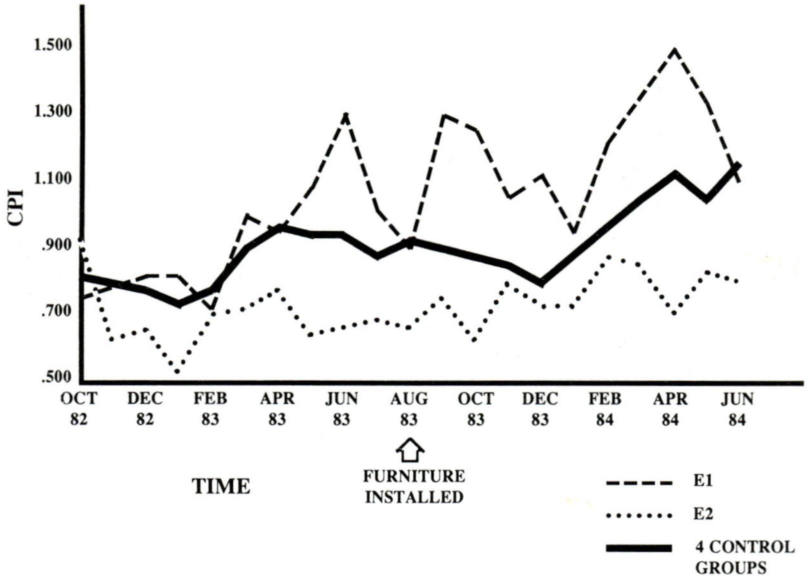

Figure 1.5. Combined productivity index of experimental and control groups between October 1982 and June 1984.

Other features which received significantly improved satisfaction ratings by both experimental groups included: furniture adaptability and arrangement, worksurface space and arrangement, availability of conference facilities, and electrical outlets. One control group was significantly less satisfied with the worksurface height ($p<0\cdot05$) and another control group showed dissatisfaction with the amount of filing space in the workstation ($p<0\cdot01$).

Discussion

Costs/benefits

The systems furniture workstations required less space than the improved conventional furniture to provide the required worksurface and storage space. The total space savings realized by the systems furniture work group as compared with the improved conventional work group was 380 ft^2 (1750 – 1370 ft^2). At the time of the study, the average cost of leased office space was $9.0 per square foot in the area immediately surrounding the government agency's location. In terms of leased space cost avoidance, savings of $3420 were realized by comparing the space efficiency of the two furniture conditions.

In addition, labour savings can be calculated based on the 20·6% increase in performance demonstrated by the workers in the systems furniture work group. The

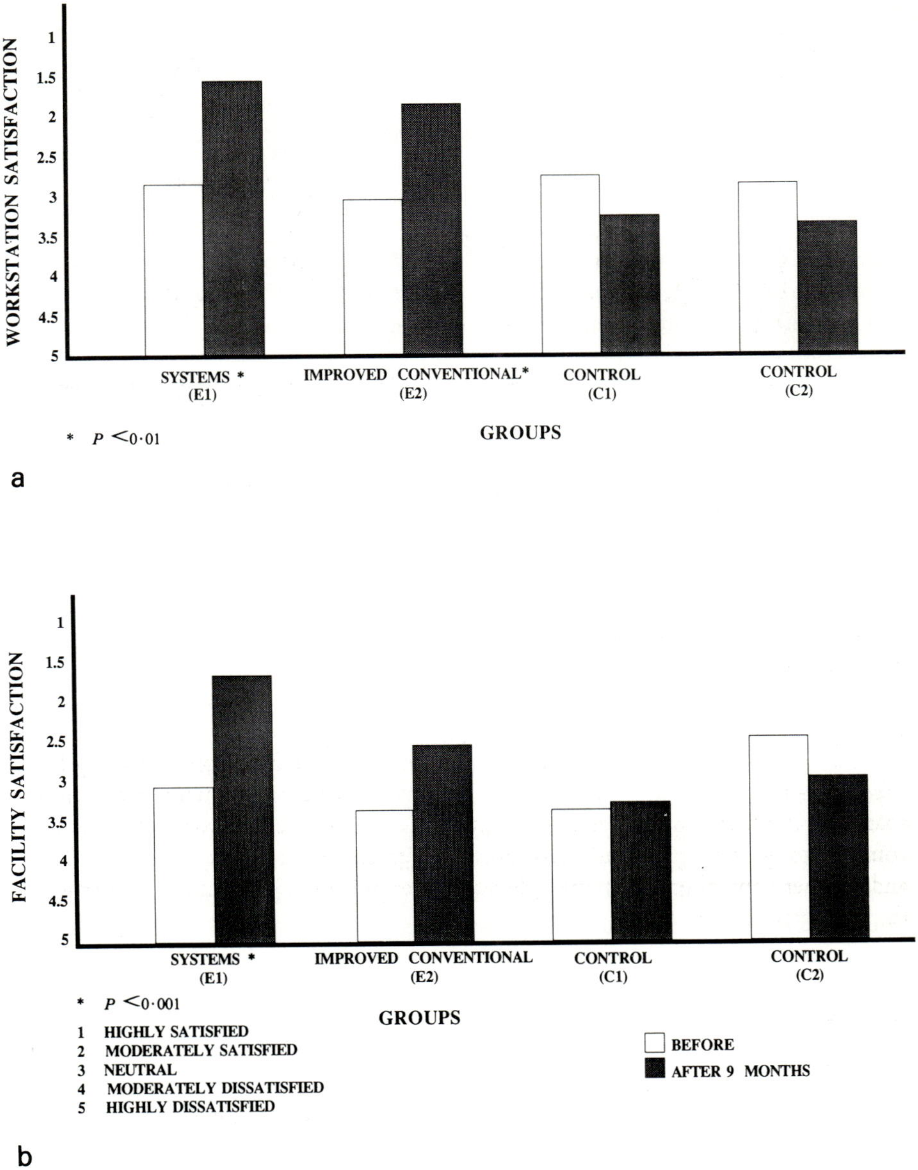

Figure 1.6. Mean satisfaction with the (a) workstation and (b) office facility before and 9 months after furniture installation.

annual labour cost of a procurement agent in this study was $19 735. The annual labour savings realized from a 20·6% performance increase is $4065 per worker ($56 914 for a work group of 14 individuals). When the total labour and space savings of the systems furniture group ($60 980) was compared to the total cost of the systems furniture ($54 509), a simple payback of 10·8 months was calculated.

Conclusions

It can be concluded from this study and others that improving the workspace by providing comfortable, adjustable furniture functionally arranged to satisfy individual and organizational office requirements can facilitate employee performance and satisfaction with the workplace.

Efforts to improve the workspace in this study resulted in the expected pre/post increase in performance and satisfaction for the systems furniture group (E1), but did not result in improved performance for the improved conventional work group (E2). Performance for E1 improved by 20% as compared with the control group which showed no significant change in performance. A simple payback calculated from labour and space savings revealed a payback on the systems furniture investment in 10·8 months.

Since every effort was made to control all other factors between the groups except the type of furniture, these results indicate that the type of workstation furniture provided is an important factor in improving the workspace. Workstation design and layout were developed on the basis of a thorough understanding of employee and work group operations and facility requirements in the present study. Deficiencies of the existing workspaces were identified and corrected in the new design. The degree to which the workstation design and layout successfully satisfied the human and functional office requirements affected performance in the present study. Many other environmental factors which were not manipulated in the present study have also been shown to be related to worker performance and office satisfaction, and should be considered in efforts to improve the workspace.

Results from the office satisfaction questionnaire indicated that E2 reported an overall increase in satisfaction with the workstation, but rated the office facility as 'neutral', in general, both before and after furniture was installed. Items listed in the office facility section of the questionnaire included both features which were manipulated in the study and features which were not directly manipulated, but may have been affected by the addition of workstation partitions and task lighting, such as ambient lighting, lighting brightness level, type of lighting, ease of building circulation, and the heating, ventilation and air conditioning system. E1 reported significantly improved satisfaction with over twice as many questionnaire items as did E2, including those office facility items which were affected by the experimental manipulation.

The control group registered no significant increases in satisfaction over the course of the study; instead, the group reported significantly lower satisfaction with some features, such as worksurface height and amount of filing space following the furniture installation in the experimental groups. This result may be due to the proximity of the control groups to the experimental groups which received improved workspaces. Greater dissatisfaction can result as employees compare their workspaces with workspaces perceived to be 'better'. From this finding it can be speculated that with the wide marketing of ergonomic furniture and increasing frequency of application of this type of furniture in everyday offices, the typical worker is cognizant of its existence and possesses expectations for comfort which were not as apparent several years ago — a serendipitous, but perhaps valuable, finding.

Limitations

The sponsoring agency anticipated that computerization software under development would become available shortly after furniture installation. The new workstations were designed and arranged to permit sharing of computer terminals between every two procurement agents. However, computer terminals were not installed in the workstations during the course of this study because of delays in software development. The study was ended in June 1984 because agency management could not further postpone their plans for reorganization of the units under a new management programme. Workers were to be reassigned to different work units and the systems furniture was 'internally procured' by a higher-ranking work unit of software engineers who wanted the furniture for their own use. It is, therefore, not known what impact the addition of computer terminals in the new workstations would have produced on worker performance and satisfaction.

In interpreting the results of this study, it should be noted that the number of subjects was small (total $n = 74$, including control groups). The variability in the performance data between months, although not uncommon in field studies, cannot be explained through statistical analysis or from discussions with the agency's management. The researchers frequently contacted management to identify any changes which had occurred in policy or operations during the course of the study. Although minor changes were documented, it was the belief of agency management that their effects on performance would be very limited.

A positive significant correlation existed between backlog and performance when taken across all groups for the months before furniture installation. The correlation was not significant over all groups for the months following furniture installation, however, the amount of backlog significantly increased for all groups (except E2) from before to after. Therefore, some of the variability of performance within groups for the months before furniture installation may have been due to varying backlog, which may have acted either to inflate or underestimate the amount of change over time.

It is important to use a multi-method data collection approach which combines objective and self-report methods in an effort to obtain valid and meaningful measures of performance. Every effort to identify and control for extraneous variables should be made before any intervening action is taken. The behaviour under study should be observed over a sufficient period in an effort to reduce the effect of variability.

References

American Productivity Center, 1982, *White Collar Productivity: The National Challenge* (Grand Rapids, MI: Steelcase, Inc.).

Brill, M., Margulis, S. T. and Konar, E. (BOSTI), 1984, *Using Office Design to Increase Productivity*, Vol. 1 (Buffalo NY: Workplace Design and Productivity, Inc.).

Dainoff, M. J., Fraser, L. and Taylor, B. J., 1982, Visual, musculoskeletal, and performance differences between good and poor VDT workstations: preliminary findings. *Proceedings of the Human Factors Society 26th Annual Meeting* edited by R. Edwards, p. 144 (Santa Monica, CA: Human Factors Society).

Dressel, D. L. and Francis, J., 1987, Office productivity: contributions of the workstation. *Behaviour and Information and Technology*, **6**(3), 279–284.

Francis, J., Dressel, D. L., MacArthur, S. A. and Neathammer, R. D., 1986, *Office Productivity: Contributions of the Physical Setting*. Technical Report P-86/13 (Champaign, IL: US Army Corps of Engineers Construction Engineering Research Laboratory).

Springer, T. J., 1986, *Improving Productivity in the Workplace: Reports from the Field* (St Charles, IL: Springer Associates, Inc.).

Sundstrom, E. D., 1986, *Work Places: The Psychology of the Physical Environment in Offices and Factories* (Cambridge: Cambridge University Press).

Westinghouse Architectural Systems Division, 1982, *A Study of White Collar Productivity and Open Office System Furnishings: An Executive Summary* (Grand Rapids, MI: Westinghouse Electric Corporation).

Wineman, J. D., 1986, *Behavioral Issues in Office Design* (New York: Van Nostrand Reinhold).

Chapter 2

Ergonomic Intervention for Better Health and Productivity: Two Case Studies

CHOON-NAM ONG

Introduction

Visual display units (VDUs) have now become commonplace in offices, banks, industries and homes. In many large organizations, such as airlines and banks, the operation of computer terminals is now a full-time job for many workers.

The development of VDUs and their incorporation into the work environment has been so rapid that some confusion has arisen. Fitting the latest technology into our work environment has become an important function of management and workers.

VDU operators remaining at the same workstation for long periods attending to a specific task were found to be more prone to feelings of tiredness (Hultgren and Knave, 1974; Smith *et al.*, 1987a). Moreover, numerous studies have suggested that a poorly designed VDU workplace can cause a variety of physical problems, including visual fatigue, stress and musculoskeletal complaints (Ostberg, 1980; Hunting *et al.*, 1981).

This chapter illustrates two case studies where ergonomics has helped to reduce the physical fatigue and improve the performance of workers using VDUs.

The first investigation was conducted in a computer centre where operators were involved mainly in data entry. This group of operators spent more than 85% of their working hours at the computer terminal. Their work was routine and was performed at high speed. An intervention programme, involving a redesign of the workplace and rescheduling of work activities was implemented. It was observed that these changes led to a reduction in physical fatigue and improvement in the work performance.

The second study was conducted in a typesetting and proof-reading firm with the aim of creating a work environment in which jobs could be performed with minimal physical stress, less fatigue and greater interest and enthusiasm. It was hoped that this improvement would help to raise work morale and productivity. The results showed that after the intervention there was a marked reduction of musculoskeletal complaints. It was also observed that when typesetters were asked to perform proof-reading work in between VDU typing tasks, as opposed to continuous VDU work, they showed a marked improvement in the quality of their typesetting work, i.e. a reduction in error rate.

Case 1

Statement of the problem

The work of this group of data-entry operators was fairly monotonous. The task was repetitive and on many occasions the workload could be very intense. Although medical examination did not reveal any statistically significant differences between the data-entry operators and conventional office workers, an earlier study had indicated that these data-entry operators had more health complaints than the conventional office workers (Ong *et al.*, 1981). The data-entry operators also reported to be more fatigued after work than the conventional office workers.

Assessment of the workplace indicated that basic ergonomic considerations had been seriously overlooked.

During the investigation attempts were made to change the physical environment and working conditions with the aim of enhancing productivity and reducing job stress among the operators.

Conditions investigated

The following aspects of the environment were studied:

1. *Workplace design* — Muscular discomfort and general fatigue were believed to be the result of poor ergonomic working postures caused by inadequate consideration of the workplace and poor layout of work items and equipment.
2. *Lighting* — Provision of lighting, both in terms of the quality and quantity of illumination.
3. *Colour scheme* — Examination of various colour schemes in the existing workplace.
4. *Noise and thermal environment* — Evaluation of the noise level with a view to reducing some of its nuisance effects; thermal comfort, ambient temperature and humidity were also examined.
5. *Working hours and rest pauses* — Determination of duration of work and frequency of rest pauses.
6. *Performance* — Keystroke rate of operators.

Observations

Workplace design

In this study, the following areas were examined:

(a) working height and working area,
(b) seating height,
(c) leg room, and
(d) visual distance.

As in many traditional offices, working height was determined by the desk height.

There were two types of working desks in this data processing centre, with heights of 70 cm and 68 cm, respectively. From a questionnaire survey, over 75% of the operators were happy with the height of the work desk.

The seating height ranged from 38 to 44 cm. From the questionnaire survey, about 60% of the female operators felt that the chairs were uncomfortable, not steady and too hard. The chairs were provided with adjustable backrests but most of the operators did not lean against the back rest.

Lighting

The illumination levels in the data room ranged from 65 to 200 lx on the source materials, and illumination on the keyboard ranged from 45 to 250 lx. On some desks, the levels were as low as 60–70 lx. In addition, there was excessive contrast in the visual field. Surprisingly, most of the operators felt that the illuminations were just right. On the other hand, over 65% of them complained about eye-strain and visual problems.

Colour scheme

The walls of the room were dark and dim. This was one of the causes of excessive contrast in visual field. The survey results also indicated that the room was considered by the users to be dull and gloomy and none thought that it was restful and friendly.

Thermal comfort and ventilation

Many operators have indicated in the questionnaire survey that certain locations were very cold and draughty. However, results also showed that the room was reasonably well ventilated. The corrected effective temperature (CET) was around 19–20°C, which was too cold for this type of sedentary work.

Working hours and rest pauses

On a normal work day, the data-entry operators worked from 8.30 a.m. to 4.30 p.m. daily. However, on days when they were required to work overtime, they had to work from 8.30 a.m. to 7.30 p.m. For the extended days, the total working time totalled 11 h with about 95 min of break (with a lunch break of 40 min and a rest break of 55 min). There was an official break of 15 min for the overtime duration, but the operators preferred to sacrifice the 15-min break in exchange for finishing work at 7.15 p.m. instead of 7.30 p.m. The percentage of rest time was about 10–15% of the total work time.

Performance of the operators

The performance of the operators, as measured by key strokes/hour, was highest in the morning, dropped by 10% in the afternoon and further when they had to perform overtime duty.

Interventions

1. *Workplace* — Work desks with new dimensions were recommended. Owing to the smaller body build of the Singapore operators, a sitting height of 46 cm and working height of 74 cm were recommended (for details see Ong *et al.*, 1988). Foot rests and document holders were provided for all operators.
2. *Lighting* — A 450–500 lx general illumination level was recommended.
3. *Colour rendering* — Ceiling, walls and floor with a reflectivity of 100%, 75% and 30%, respectively, were recommended.
4. *Readjustment of working hours* — A total of 110 min break time was recommended for a normal work day. The overall working time was 380 min with breaks totalling 23%. For extended days with overtime activities, an extra 25 min break time was recommended.

Results

Figure 2.1 shows the prevalence of discomfort in the same group of operators before and 1 year after the intervention. There was an obvious reduction of musculoskeletal complaints among the data-entry operators at the new workplace. The prevalence of visual fatigue had also reduced significantly from over 90% to 62% for those who complained about visual problems on an occasional basis. It is, however, important to note that this value is still higher than for conventional office workers (usually around 40%) (Ong, 1984).

With regard to the performance of the operators, Table 2.1 shows that the average keying speed increased from 9480 keystrokes/h (the average for 12 months before intervention) to about 13 000 keystrokes/h (the average for 12 months after the change). A more recent assessment (6 years after the physical and organizational interventions) indicated that the average speed for 20 operators was approximately 11 300 keystrokes/h (Table 2.1).

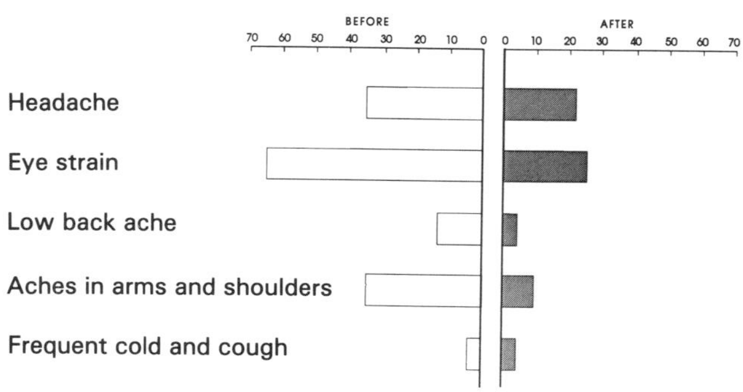

Figure 2.1 *Subjective health complaints of data-entry operators before and after the intervention. The questionnaire on subjective symptoms was adapted from Ong (1984).*

Table 2.1 Performance and errors of data-entry operators.

	1980	1982	1987
Speed (keystrokes/h)	7166–14000	12700–13392	10450–13110
Error (%)	0·19–4·73	0·10–0·13	0·18–1·10

Case 2

The second study was conducted in a local, medium-sized typesetting firm employing about 80 staff. There were 48 workers using typesetting machines for their routine work, and 16 female workers proof-reading. The company worked from 8 a.m. to 10 p.m. with two groups of shift workers.

Statement of the problem

The main task of the typesetters was keying in text through VDU keyboards to the computerized typesetting machine. Their average keying speed was about 8000 keystrokes/h.

The management noted that there was a high workforce turnover and the workers were dissatisfied with the work environment and working conditions. In addition, the poor work performance of the typesetters and proof-readers resulted in frequent back-tracking of jobs as they moved from operation to operation. Customers frequently sent back semi-finished products for rectification or rework because of critical errors and omissions in the text which upset production schedules and made production planning and control difficult and unreliable.

The purpose of this study was to identify the main causes of poor work performance and to make recommendations for improvement. It was hoped that the typesetters and proof-readers could thus deliver jobs of consistently good quality at each stage of the operation, thereby facilitating production planning.

Conditions investigated and observations

Excessive desk height was noted for operators using the imported typesetting machines. This resulted in the operator's elbow being lodged below the working height. The machine was 75 cm above the floor level. Document stands were used by this group of operators. However conventional office chairs of about 40 cm seating height were provided. They were adjustable and the chair's backrest could not be used for proper support of the lower back of the typesetters. A low illumination level was a common feature in all of the three workrooms. The maximum illumination was 260 lx. At some work desks the lighting levels were as low as 60 lx. Excessive colour contrasts in illumination levels were also noted. Obviously, there was a crucial need for changes in the workplace.

Although basic training on the use of the typsetting machine was given to each operator, they were usually given tasks immediately after a brief training session.

Based on a questionnaire survey, work in the office was considered to be monotonous by most of the typesetters.

Two short breaks of 15 min were provided throughout the whole work shift in addition to a 1-h lunch break.

Interventions

Physical intervention

The workplace was redesigned and physical changes made in all the work areas. These included provision of a new seat adjustable from 40 to 48 cm in height and a foot rest for operators who were less than 155 cm tall. Supplementary lighting was furnished for each workstation. The detailed workplace recommendations were based on a study on the relationship between Asian anthropometrics and their display station preferences (Ong *et al.*, 1988).

Flow grouping of jobs for the work team

Working in teams enabled employees to see the whole work process and to appreciate their own contributions. Typesetters were encouraged to form working cohorts with the proof-readers and production planners.

The advantages of working in teams are as follows:

1. easier production planning and control,
2. increased flexibility,
3. enlarged jobs,
4. direct feedback of results, and
5. improved human relations through team work.

Working hours and rest pauses

Typesetting work is considered as light physical work requiring a heavy mental effort. The job demands frequent rest pauses of short duration in order to:

1. prevent fatigue,
2. allow opportunities for refreshment, and
3. provide occasions for social contacts.

An extra break of 10 min for the afternoon session was recommended. This leads to a total break time of 100 min (including 60 min for a lunch break).

In order to assess the intervention programme, the performance of 35 typesetters was studied for a period of over 30 months. The subjective symptoms of these operators were also assessed before changes were carried out (1984) at the workplace and about 30 months after intervention (late 1986). Each operator was required to complete a questionnaire concerning subjective symptoms, working conditions, job stress and various aspects related to the VDU workstation. The performance of the operators was determined by both the average keying speed (in keystroke/h) and keying errors (percentage). These were assessed and recorded daily by the computer.

Results

Figure 2.2(a) and (b) show the prevalence of subjective health complaints for 1984 and 1986. There was an obvious reduction of musculoskeletal complaints. A relatively smaller proportion of the operators claimed to have musculoskeletal fatigue in the new workplace compared with that at the old office.

As shown in Figure 2.2(a), the prevalence of subjective symptoms such as 'miserable and unhappy', 'tiredness without knowing why' was also reduced. The prevalence of visual fatigue was also reduced from over 60% to about 20% (Figure 2.2 (b)).

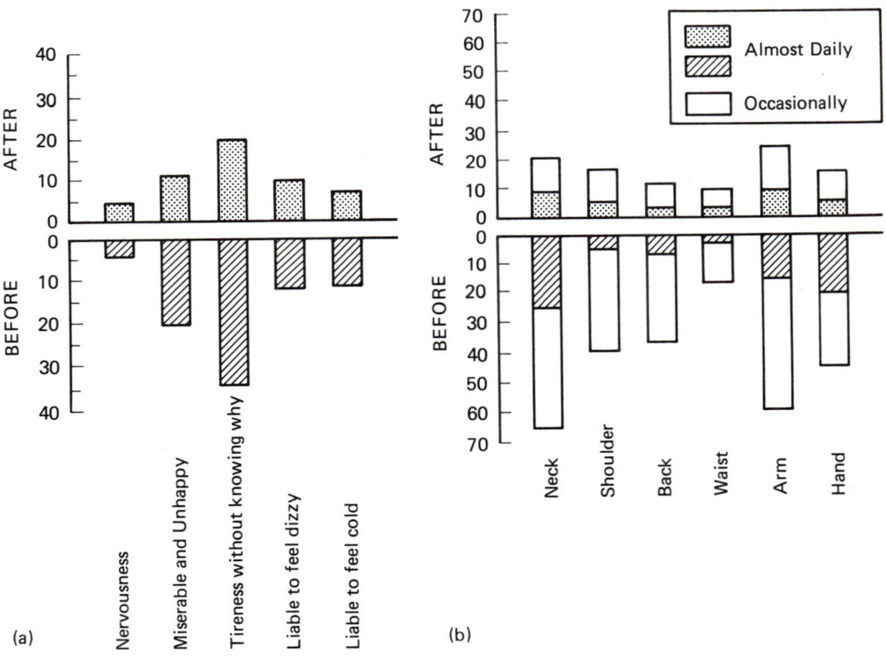

Figure 2.2 Subjective health complaints of typesetters before and after intervention. The questionnaire on subjective symptoms was adapted from Ong (1984).

Performance and errors

With regard to the performance of workers, Table 2.2 shows that the average keying speed had decreased from 7567 keystrokes/h (average for 12 months) to 7420 keystrokes/h for the young operators. It was, however, noted that for the older VDU operators the drop in keying speed was more significant. A 7% decrease in speed was observed. On the other hand, there was a significant decrease in error rate from around 0.08% to 0.04%.

Following the intervention, the management reported a marked decrease in labour turnover which was maintained for the following 2 years. The typesetters also reflected higher morale compared with that prevalent before the changes although quantitative data were not available.

These results show that a better working environment is conducive to higher job performance and reduced health complaints.

Table 2.2 Performance and errors of typesetters.

	1984	1986
Speed (keystrokes/h)		
Average	7567	7420
<40 years old	8105	7985
>40 years old	7240	6786
Errors (per 10 000 characters)		
Average	7·8	4·0
<40 years old	6·5	2·5
>40 years old	9·6	6·1

Discussion

Introduction of new technology often leads to an increase of efficiency, but sometimes may also give rise to adjustment problems for the users. Various studies have indicated that neglecting ergonomics/human factors in VDU work gives rise to adverse health effects and performance decrements (see review by Scalet, 1987). On the other hand,it is believed that improved work organization and environment leads to a higher worker performance. Both studies reported here have indicated positive developments after changes in the workplace and work organization.

Physical intervention and job design

The work environment comprises both physical environment (workplace design, lighting, noise, colour scheme, etc.) and job organization (working hours, supervision, motivation, etc.). The original aim of the first study was to improve the physical work environment and to overcome some of the apprehensions regarding the health aspects of VDU work. It was, however, believed that the problems at the VDU workplace could not be dealt with by technical means alone. Organizational modifications would also be necessary to improve the quality of working life. Although changes carried out at the VDU workplace were mainly focused on the physical work environment, various improvements in the organizational aspects were also investigated. The positive results on the work performance after the changes were unlikely to be a consequence only of the physical changes. The modification of the work–rest schedule and other changes could also have contributed to the improvements.

Anthropometrics and VDU workplace

Knowledge of anthropometric dimensions is an important requisite for the design of equipment and workplace. In Singapore, a large proportion of VDU workstations are imported from overseas. These imported VDU workstations have been designed for the Caucasian physique and hence have posed numerous musculoskeletal problems for the local operators who are of different anthropometric dimensions from their western counterparts (Ong *et al.*, 1988). There is also a growing recognition that postural preferences are badly neglected in most of the workplaces. Based on an earlier study of the anthropometrics of 104 VDU operators of three ethnic groups, it was found that owing to a smaller body build, Singapore operators preferred to have a sitting height of about 46 cm, and a working height of about 74 cm, compared with 47 cm and 77 cm, respectively, as preferred by their European counterparts. A questionnaire survey after the ergonomic intervention indicated that the majority of the typesetters were happy with the new workplace and the work items after the changes.

Work–rest schedule

Full-time work at a VDU is demanding, both visually and psychologically (Smith *et al.*, 1987b). The operator must sit for long periods in what may not be a fully comfortable position; the eyes are continually under at least some degree of strain; and prolonged periods of intense concentration are necessary in order to execute assignments that are often mechanical and less than interesting.

As regards the work–rest schedules for VDU operators, there are some general recommendations (e.g. 15 min rest after 2 h of continuous work), but evidence supporting this recommendation has not been found (Hendricks *et al.*, 1986). The National Institute for Occupational Safety and Health (NIOSH) has recommended a 15-min rest break after 2 h of continuous work for operators under moderate visual demands and/or moderate workload. It was also mentioned that the period of continuous work should be reduced to 1 h for operators under high visual demands, or high workload, or for those operators engaged in repetitive work tasks. However, 'moderate' and 'high' workload were not defined by NIOSH (1981). There is also a lack of experimental support from available literature for an appropriate work–rest scheme for VDU work.

In data-entry work, the task is monotonous and repetitive, and it is often carried out on a full-time basis (Johansson, 1986). It contains many features of short-cycle repetitive work, rigid work procedures, high production standards and constant pressure on performance; these operators require vigilance. A well-designed work–rest regime would help to keep them in an alert and fresh state. Short recess times would help to minimize the visual and muscular fatigue of the operators. The results shown here indicate that provision of more rest pauses did not cause any deleterious effects on the performance, but speeded up the work in case 1, and improved work quality in both cases (Tables 2.1 and 2.2). More research on how rest breaks affect performance is obviously needed.

Conclusions

The scope of application of ergonomics in the VDU work environment is tremendous. It was noted in both cases reported here that improvement in the working environment with ergonomic considerations led to significantly less visual and muscular discomforts. In addition to the reduction of physical fatigue, the efficiency and performance of the operators improved markedly (for the first case). With an increasing number of businesses shifting to computerization, the crucial need for changes in office design and work scheduling should gain wider recognition. There is unlikely to be any harm in improving the ergonomics of the workplace and it may help to solve inherent health problems to a great extent. If it does not, it will probably be good for productivity.

Acknowledgements

The author is grateful to Ms S. L. Yeong and Ms W. C. Kee for their valuable criticism of the manuscripts. I also thank the management and staff of Singapore Airlines for their assistance in this study.

References

Hendricks, D., Kilduff, P., Brroks, P., Marshak, R. and Doyle, B., 1986, *Human Engineering Guidelines for Management Information System* (Maryland: US Army Human Engineering Laboratory).

Hultgren, G. V. and Knave, B., 1974, Discomfort glare and disturbances from light reflections in an office with CRT display terminals. *Applied Ergonomics*, **5**, 2–8.

Hunting, W., Laubli, T. H. and Grandjean, E., 1981, Postural and visual loads at VDT workplace: I. Constrained postures. *Ergonomics*, **24**, 917–923.

Johansson, G., 1986, Growth and challange vs wear and tear of humans in computer mediated work. *Proceedings of the International Scientific Conference: Work with Display Units* (Stockholm) pp. 249–251.

NIOSH, 1981, *Potential Health Hazards of VDTs*. Publication No. 81–129 (Washington, DC: DHHS/NIOSH).

Ong, C. N., 1984, VDT workplace design and physical fatigue: a case study in Singapore. In *Ergonomics and Health in Modern Offices*, edited by E. Grandjean, (London: Taylor and Francis) pp. 484–494.

Ong, C. N., Hoong, B. T. and Phoon, W. O., 1981, Visual and muscular fatigue in operators using visual display terminals. *Journal of Human Ergology*, **10**, 48–53.

Ong, C. N., Koh, D., Phoon, W. O. and Low, A., 1988, Anthropometrics and display station preferences of VDU operators. *Ergonomics*, **31**, 337–347.

Ostberg, O., 1980, Accommodation and visual fatigue in display work. *Displays*, **1**, 81–85.

Scalet, E. A., 1987, *VDT Health and Safety: Issues and Solutions* (London: Ergosyst Associates).

Smith, M. J., Carayon, P. and Miezio, K., 1987a, VDT technology: Psychological and stress concerns. In *Work with Display Units 86*, edited by B. Knave and P.-G. Widebeck (Amsterdam: Elsevier) pp. 695–699.

Smith, M. J., Cohen, G. F., Stammerjohn, L. W. Jr. and Happ, A., 1987b, An investigation of health complaints and job stress in video display operations. *Human Factors*, **23**, 387–400.

Chapter 3

Employee Comfort, Satisfaction and Productivity: Recent Efforts at Aetna

CAROL SULLIVAN

Introduction

Background

Aetna Life and Casualty, based in Hartford, Connecticut, USA, is the nation's largest investor-owned insurance and financial services institution. The company employs over 43 000 workers nationwide. Assisting these employees are 18 large-scale computer systems, processing over 21 million transactions per day and supporting a network of about 24 000 visual display terminals (VDTs), including word processors and personal computers (PCs). Aetna currently has a terminal-to-worker ratio greater than 1:2.

Facilities

Almost 19 000 Aetna corporate headquarters employees currently occupy numerous owned and leased facility complexes in and around Hartford. Two of these facilities — the Rogers Building, completed in 1929, and the Tower Building, completed in 1972 — are currently undergoing major renovation.The majority of the approximately 5000 employees who occupy these two facilities are from two of the major operating divisions: Commercial Insurance and Personal Financial Security.

The construction of the Tower Building signalled Aetna's full commitment to electronic data processing and the major role it was to have in work at Aetna. The building incorporated centralized data processing and unobstructed, flexible office areas, as well as windowless workspace to control energy and light. However, this monumental concrete space with its dark wood panelling was intimidating and alienating to many employees.

Unrenovated space in both buildings is 'open' but interspersed with columns and many partial wood and drywall enclosures, which sprang up in response to needs for more privacy and less distraction. They house officers (management-level employees) and certain non-officer professional employees, as well as data processing (DP) equipment, but cut off natural light and create problems in air and traffic circulation. Non-officers, who include administrative (clerical) and technical/professional personnel in unrenovated areas, occupy traditional metal desks and chairs, while officers occupy wood desks and chairs. Filing cabinets and numerous pieces of mix-and-match furniture to accommodate equipment and project work also fill the unrenovated space. These problems and the solutions provided by the renovation are discussed in detail later in this chapter.

Employees

Aetna's Statement of Principles lists its employees as its greatest assets, stating that their success is the most critical factor in achieving its mission. One of the ways Aetna attempts to ensure their success is by providing an environment, including facilities and technology, that give optimal support to employees for the work they do. Various groups within Aetna are concerned with how well the environment supports employees, and these groups frequently work together on multidisciplinary teams. Aetna's Facilities Planning Department is responsible for planning, building and outfitting owned and leased Home Office facilities. Most of the survey work on employee environmental satisfaction is done by this group.

Others who participate in these teams include:

1. facilities management and operations professionals,
2. technology and telecommunications planners,
3. those who monitor life safety,
4. human resource professionals, concerned with issues ranging from organizational effectiveness to employee health,
5. 'People/Technology Programs', a special group created in 1981 to assess the impact of technology on Aetna employees in areas such as hardware and software usability and health issues, including lighting, ventilation and furniture, as well as job and environmental stress, and
6. engineers, whose major role is to investigate commercial insurance risks and claims.

Most importantly, Aetna employees play an active role by participating in planning their department space, assisting in co-ordinating space moves and evaluating the facilities by responding to surveys.

Research orientation

For almost 10 years, in recognition of the fact that people, equipment and environmental support options are continually changing, Aetna has been the site of a real-life research laboratory to study these changes. Researchers and employees have co-operated in a number of projects (studies and interventions) to improve environmental conditions and worker comfort, and to provide a more flexible space for change. These projects have taken a five-step approach:

1. evaluation of existing space,
2. definition of present and future functional needs,
3. testing and evaluation of products in real-life work settings,

4. selection and installation of products that satisfy the greatest number of needs, and
5. evaluation of selected products by end-users.

The methodology used in this research varies somewhat among studies, but has included:

1. live-in mockups with furniture, carpeting, lighting, power distribution, heating, ventilation and air conditioning (HVAC) and window treatment options,
2. questionnaires and interviews, with numerically scaled and open-ended questions about job activities, comfort, satisfaction with environmental features and the effect of these features on productivity,
3. group discussions with employees, and
4. physical environment measurements.

This chapter briefly describes the studies and interventions that have occurred over the past 10 years and discusses two recent studies in greater depth.

Studies and interventions

The studies and interventions which have influenced environmental change at Aetna in the past 10 years are as follows:

1979 — *DP Workstation Study*
Result: Replacement of traditional, metal furniture with systems furniture in all Home Office data processing departments.

1983 — *Functional Office Study (FOS)*
Result: Positive feedback from users of systems furniture and negative feedback from users of traditional furniture in areas of comfort, functionality and privacy led to a decision to give systems furniture to all non-officers.

1983–5 — *Middletown Intervention*
A new headquarters facility, constructed for the Employee Benefits Division, was the first to phase-in systems furniture for its employees, based on the recommendations of the FOS study.

1985 — *Middletown Post Occupancy Evaluation (POE)*
Input from this evaluation influenced refinements to this facility and decisions made in planning the Home Office Renovation (the renovation of the Rogers and Tower Buildings).

1985–91 — *Home Office Renovation*
Two facilities (the Rogers and Tower Buildings) which house Home Office employees from the Commercial Insurance and Personal Financial Security Divisions are currently being renovated. All traditional metal furniture is being replaced with systems furniture, based on the recommendations of the FOS study. This renovation is described in greater detail later in this chapter.

1985 — *Home Office Renovation Mock-up*
Four mock-ups totalling over 17 000 ft^2 were constructed, lived-in and evaluated by employees and by in-house experts. Result: Decisions were confirmed and refined in areas such as choice of systems furniture, VDT accessories, life-safety features, improvements to the HVAC system, installation of a sound-masking system and adoption of a task-ambient lighting system.

1985 — *Officer Furniture Study*
In connection with the renovation, this study was undertaken to provide officers with furniture that would fit smaller private offices, and better accommodate VDT equipment. Two wood furniture lines were chosen to meet these needs, based on officer evaluations.

1985 — *Chair Study*
This study was also conducted as part of the renovation planning process. Employees and experts worked together to identify the functional and ergonomic characteristics needed from chairs, and to evaluate a variety of chairs. Result: Several ergonomic chairs were chosen for the renovation and are also being introduced in other Aetna facilities, including field offices.

1987 — *Middletown — Second Post Occupancy Evaluation*
In order to test refinements that had been made to this facility and to provide further input to the Home Office Renovation planning process, a second evaluation was conducted. Feedback from this survey confirmed the importance of careful attention to HVAC comfort, acoustical distraction and meeting and project space.

1987 — *Home Office Renovation Post Occupancy Evaluation*
This evaluation was conducted during the first year of the 5-year renovation period, in order to evaluate certain features of the renovation and correct problems where possible. Results of this evaluation are discussed later in this report.

1986–9 — *Employee Productivity and Satisfaction Study (Productivity Study)*
This study was undertaken as a co-operative effort between the Facilities Planning Department and a Home Office department to examine the relationship between improvements to the physical work environment (the renovation) and employee productivity and satisfaction. The results of this study are discussed in-depth later in this chapter.

Aetna's Home Office Renovation (the renovation of the Tower and Rogers Buildings) is the cumulative result of all of the above research and interventions. The remainder of this chapter describes the environmental changes brought about by the renovation, and discusses employee reactions to the renovation as elicited in two recent studies — the Home Office Renovation POE (1987) and the Productivity Study (1986–9).

Home Office Renovation

In 1986, work began on the complete revitalization of Aetna's two main headquarters facilities — the Rogers and the Tower Buildings. The master plan was conceived as a solution to the following problems:

1. *HVAC discomfort* — Pre-renovation HVAC discomfort was caused primarily by an outdated system, inadequate for increases in people, equipment and partitions. The renovation provides (i) more mechanisms to supply fresh air and pick up return air, placed to complement the new layout; (ii) individual controls for all closed offices; and (iii) an improved building automation system for better reporting and response to variations.
2. *Acoustical and visual privacy and distractions* — Pre-renovated space was crowded with people and equipment. Existing partitions were partial only, and there was little or no provision for acoustical comfort. The renovation provides (i) a sound-masking system; (ii) workstations with two to three panels (some acoustical), depending on the need for team work; and (iii) fully enclosed offices, conference rooms and DP equipment rooms.
3. *Lighting/glare discomfort* — Lighting in pre-renovated space was provided by ceiling fixtures that can cause glare on worksurfaces and on VDT monitors. The venetian-blinded windows in the Rogers Building can cause additional glare when the blinds are open, or a gloomy environment when they are closed. Natural light is often blocked by partitions. The renovation solves these problems by (i) placing windowed enclosed spaces in core areas (both closed and open areas have access to natural light); (ii) replacing venetian blinds with shades that filter glare, yet allow daylight and an outdoor view; (iii) designing workstation layouts to permit proper VDT-monitor placement in relation to light sources; (iv) recommending glare-screens for all VDT-monitors; and (v) installing a combination task/ambient lighting system to allow more individual control for a variety of tasks.
4. *Furniture discomfort* — Traditional metal desks provide limited worksurface and storage and do not accommodate VDTs. Extra furniture to provide for these needs crowds the space, is not optimally adjustable and is inaccessible. The renovation provides (i) a two-surface individual work space (L-shaped for non-officers); (ii) varying amounts of storage, depending on need; (iii) options for VDT accommodation, including adjustable keyboard and monitors arms; and (iv) improved, easy-to-adjust chairs.
5. *Crowded, unsafe space* — Personnel, equipment and furniture proliferation, compounded by frequent relocations, have resulted in a crowded and confusing individual and group workspace and layout. The renovation solves these problems with (i) a 58 ft^2 common footprint workstation for all non-officers; (ii) two standard office sizes for officers; (iii) two worksurfaces for all employees with a variety of componentry options to suit individual needs for storage and equipment accommodation; (iv) provision of alternative work and meeting spaces (conference rooms and oases); and (v) layout of department space to allow clear, unobstructed circulation. Enclosed offices, conference rooms and oases are

located in the central core with a primary circulation path dividing them from the workstation area.

Workstations are placed in rows, three to five deep, and side-by-side in groups of six to ten along central, fixed power-management panels. Multiple-use areas are placed between workstation groups for storage, equipment or conferencing.

In an effort to evaluate the success of the Home Office Renovation, several studies were conducted recently. Two studies that will be discussed here include a Home Office Renovation Post Occupancy Evaluation and a three-phase Employee Productivity and Satisfaction Study (Productivity Study).

Home Office Renovation Post Occupancy Evaluation

Background

In 1987, Facilities Planning designed a POE to examine employee satisfaction with certain features of the renovation. The concerns addressed in this POE included user adjustability, lighting/glare, equipment accommodation and storage. Employees were asked to rate features including chair adjustability, keyboard arm adjustability and comfort, task light adjustability and location, storage adequacy and VDT accessibility.

Respondents

In August 1987 surveys were sent to all employees who had moved to renovated space. A total of 593 officers (Os) and non-officers (N-Os) from two divisions had moved since March 1987, when renovated space first became available. These employees worked in 11 units (departments), in areas such as planning, business development, human resources, claim and actuarial. Employees in one division who represented 60% of the 'move-ins' were asked to report on renovated space in two floors of the Rogers Building. Employees in the other division, representing the remaining 40%, were asked to report on one floor of the Tower Building. The response rate was 63%, with exactly 60% of the respondents reporting on the Rogers Building and 40% on the Tower Building. N-Os made up 81% of the population. Os were 82% male, while N-Os were 69% female.

Method

Employees were asked to rate the list of design features according to a five-point Likert scale from 1 (very poor) to 5 (very good) — with 3 being average. In addition, Os were asked to comment on the renovation and N-Os were asked to list 'best' and 'worst' features of their workstation. Through *t*-tests and one- and two-way analysis of variance (ANOVA), ratings were compared (1) by division, (2) by O/N-O, (3) by N-Os with differing responsibilities, (4) by male/female, (5) by different location, and (6) by different lengths of time in renovated space.

Results

Five items to be rated were similar in O and N-O survey forms. Table 3.1 is a summary of the average ratings each group gave certain design features. N-Os from the division housed in the Tower Building rated a number of features significantly higher than those in the Rogers Building. Since mean ratings were uniformly high, the difference may be explained by the fact that the contrast between old and new was much greater for Tower than for Rogers occupants. ANOVA procedures indicated that length of time in renovated space affected satisfaction with certain features. Officers' chairs, and non-officers' keyboard arm comfort and storage became more popular after one week of use, while non-officer keyboard arm adjustability and task light location became less popular.

Table 3.1 Average design feature ratings.

Design feature	O	N-O
Chair adjustability	3·9	4·3*
Shelf adequacy	4·1	3·8*
Storage adequacy	4·0	3·9
Keyboard arm adjustability	3·6	3·7
VDT accessibility	3·3	3·6
Printer location	2·9	
Task light adjustability		3·7
Task light location		3·7
Keyboard arm comfort		3·6

*Indicates a significant difference ($P<0{\cdot}05$).

The lowest scores in the survey went to officer equipment accommodation (VDT accessibility and printer location). A number of officers commented that their offices were not appropriately set up for PC equipment. Part of the problem may have been the delays many officers experienced in receiving all their furniture and options. Although non-officers rated their task lights above average on adjustability and location, and frequently cited it as a 'best' feature, about one-third were dissatisfied with the limitation of having one task light that does not evenly illuminate both worksurfaces. The result is a workstation that, for some, is a contrast of glare and shadow, especially in workstations with shelves. ANOVA tests revealed that employees at the highest non-officer level were significantly less satisfied with task lights than employees at the lowest level. Frequent users of the keyboard arm were more satisfied than infrequent users.

As a result of these findings, other options have been provided to officers to accommodate equipment. Follow-up interviews reveal that officers are much more satisfied with these options. In addition, the keyboard arm has become an optional feature, and a task light study is currently in progress.

The POE elicited many positive comments about the renovation, and above average scores for most features. Non-officers at all levels of responsibility most frequently mentioned 'amount of worksurface,' 'increased privacy and less distraction', and the 'chair' as 'best' features of their workstations. Officers commented on the 'significantly improved working environment', and 'lower noise level . . . contributing to increased productivity'.

Employee Productivity and Satisfaction Study

Background

In 1986, a study team was formed and a multi-method research approach was developed to examine the effects of the environmental changes connected with Aetna's Home Office Renovation on employee productivity and satisfaction over 3 years and three phases of change.

The three phases included (1) old furniture in unrenovated space, (2) new furniture and layout in unrenovated space, and (3) new furniture in renovated space. At each phase, departmental productivity trends were examined and employees' attitudes about physical space were polled through a survey.

Respondents

The departmental unit under study served as a liaison between Home and Field Offices, processing requests from the field pertaining to (1) producers' (salespeople's) licenses to sell insurance, (2) their contracts with the company, and (3) their commissions from selling insurance. The number of employees in this unit varied from 43 at Phase I to 36 at Phase III. Despite gains and losses of employees throughout the study, there was a core group of 28 people who filled out the attitude survey at the end of each of the phases. These 28 included the manager who headed the unit, two supervisors and 25 technicians, later called Customer Service Representatives. Unit employees ranged in age from 20 to 55 years, with 37 being average. They had worked at Aetna for an average of 9·5 years, and in the department for 6·5; 88% were female.

The research method

The productivity analysis

In analysing productivity trends over the three phases, the team made use of weekly productivity reports that the unit kept. These reports provided a detailed list of the amount of work processed by the unit during that week (productivity or output), and the amount of work that did not get processed and had to be carried over to the next week (carryover). The team also monitored attendance (including absenteeism and turnover), as well as overtime, during the three-phase study, to determine the number of employee hours spent in performing the work (input). A ratio of total production (output) to hours worked (input) was calculated for each week and for each phase, and the phases were compared using multiple regression. In addition, the correlation between the size of the work backlog and productivity was examined.

The attitude survey

The attitude survey was administered three times to the employees — at the end of each of the three phases. It had been pilot-tested by a department whose work was similar to the one being studied. The questions in the survey were based on information obtained from employee interviews and from behavioural observations in the unit. Workspace and productivity literature was also reviewed.

In the survey, four and five-point Likert scales were used to measure employees' reactions to features of the physical work environment. The average (mean) employee responses to the survey questions were then compared as follows:

(a) Phase I with Phase II,
(b) Phase II with Phase III, and
(c) Phase I with Phase III.

Paired comparison *t*-tests were used to examine significant differences between the responses of the 28 core people. The responses of the entire unit at all three phases were also analysed. In this report we have only included average responses of the 28 core people, since they present a more accurate comparison of changes in employee attitudes about the physical environment. Regardless of whether survey questions were positively or negatively worded, the average responses in the tables in this report have been recoded so that the best score is always a 4 or 5 and the worst score is a 1.

The survey covered a number of issues that were major themes in previous studies and interventions, including:

1. space adequacy (personal workspace, departmental/meeting space, circulation space),
2. access (to people, equipment and resources),
3. comfort (furniture, equipment, HVAC and lighting),
4. privacy (acoustical and visual privacy and distraction), and
5. influence of physical space on productivity.

SPACE

Space in work settings has been discussed in terms of the amount of floor area available to individuals and groups to provide order, functionality, comfort and status (Becker, 1981; Brill *et al.*, 1984; Sundstrom, 1986). Comfort is discussed below. Status was examined in this study, but is not covered in this report. We hypothesized that for this group, a functional space would provide adequate amounts of individual and group workspace, including worksurface and storage as well as space supportive of job activities (among them reading, writing, working with equipment, telephoning, thinking, organizing, having large and small meetings, moving around the space and doing several activities at once).

ACCESS

Access to resources such as people, equipment, files, manuals, etc. can be critical factors in job productivity (Becker, 1981). Functional space arrangement locates resources in convenient places, makes it easy to get to these places, and provides appropriate places in which to use these resources.

COMFORT

Comfort in the work setting is provided by appropriate furniture, temperature, air quality and lighting. Comfort has been linked with job and environmental satisfaction, and with productivity (Louis Harris and Associates, Inc., 1980; Hedge, 1982; Brill *et al.*, 1984; Sundstrom, 1986; Wineman, 1986). This study offered an opportunity to evalute Aetna's recent efforts to improve worker comfort through furniture, lighting and HVAC improvements, and to examine the effect of comfort on productivity.

PRIVACY

Literature on privacy has discussed the concept in terms of being able to control access to oneself, to one's space and to information (Proshansky *et al.*, 1970; Westin, 1970; Altman, 1974). Achieving the right balance between privacy and interaction can be crucial to job productivity. This right balance depends on the nature of the job being performed, but also on personal factors. In many organizations, the privacy provided by environmental supports is a function of status (Laufer *et al.*, 1973; Justa and Golan, 1977). Other studies of employees at different status levels — management, professional/technical and clerical — suggest a graduated scale of types of privacy needed, based on job responsibilities. These are (1) control over access to space and self (all levels), (2) control over distractions (professional/technical and management), and (3) speech privacy (management) (Brill *et al.*, 1984; Sundstrom, 1986). We wished to test a hypothesis that employees at all levels need all three types of privacy.

RELATION BETWEEN WORK ENVIRONMENT AND PRODUCTIVITY

Previous literature has suggested a relationship between certain aspects of the work environment, job satisfaction and worker productivity (Louis Harris and Associates Inc., 1978; Becker, 1981; Sundstrom, 1986). Brill *et al.* (1984) found a direct link between supervisory and worker ratings of job performance and (1) degree of enclosure (panels or walls) and (2) layout. Many of the questions in the survey asked how effective specific environmental supports were in helping people work efficiently. Questions were also included about the overall effect of the individual workspace and the department space on people's ability to do their jobs, as well as about how specifically the overall environment affected this ability.

Interventions

Phase II

At Phase II, all non-officers' traditional metal furniture was replaced by new workstations (L-shaped, with two worksurfaces), ergonomic chairs and task lights. These workstations were equipped with one 61″ and two 53″ high, acoustical panels, various options in shelves and paperflows, and two or three pedestals for file storage. Employees who worked together in regional teams were clustered together. During Phase II, shared equipment was housed on CRT-height tables in multiple use areas between rows of workstations, and perpendicular to the windows to minimize glare. The workstation layout resulted in wider and better defined circulation paths. The manager continued to occupy his panelled area and traditional furniture. The changes introduced in the Phase II intervention addressed the following concerns:

(a) individual (workstation) space adequacy,
(b) access to resources within the individual work space, and to equipment, resources and people within the unit, due to improved layout,
(c) comfort from workstation and shared furniture that is more adjustable, from placement of VDT monitors to avoid glare and from individual lighting that employees can control, and
(d) privacy for the individual (visual and acoustical) provided by the workstation panels.

Phase III

Phase III changes included locating the manager and some oasis meeting spaces in the central core of the unit's floor area, with the workstations around the window periphery. At the manager's request, most VDTs were now placed in the workstations and shared between two people by way of carousel worksurfaces. Each such employee was given an articulating keyboard arm. Multiple use areas now housed files and some equipment. Other areas of change included indirect overhead lighting, sound-masking, HVAC improvements, glare-filtering window shades, new departmental file storage, carpeting and paint.

Phase III changes addressed the following concerns:

(a) unit space adequacy through the addition of more meeting areas,
(b) access to equipment and resources through the placement of VDTs in the workstations and of filing cabinets in the multiple use areas,
(c) comfort from keyboard arms, glare-filtering window shades, indirect overhead lighting, and a redesigned HVAC system, and
(d) privacy (acoustical) as provided by the introduction of sound-masking.

Summary of results

General productivity and satisfaction findings are organized and discussed below according to the phase of the study in which they occurred. (Responses to specific survey items and productivity effects across all three phases are summarized in Tables 3.2–3.7.) A more specific discussion of productivity and other performance effects is then presented.

Phase I — Old furniture and unrenovated space

Employee dissatisfaction with the physical environment was apparent in Phase I survey results. Almost 91% indicated that their personal work areas felt crowded and that lack of space made it difficult to do such everyday activities as organizing paperwork, reading and writing. Over 73% felt the unit did not have enough space for all the activities that should go on there and 86% said the unit felt crowded.

Table 3.2 Satisfaction with workspace.

	Phase I	Phase II	Phase III
1. Satisfaction with overall and individual unit space (5-point scale):			
My workspace has enough room for me to do an efficient job	2·39*	3·93	3.71***
The overall unit has enough space for all the activities that should go on here	2·00*	2·64**	3·48***
2. Functionality of space for performing activities (4-point scale):			
Individual workspace			
Getting in and out of my workspace	2·93*	3·61	3·79***
Reading	2·75*	3·50	3·57***
Writing	2·86*	3·54	3·68***
Organizing papers	2·54*	3·54	3·61***
Reaching for manuals/books	2·75*	3·25	3·32***
Working on project with other person	2·14*	2·92	3·19***
Shared unit space			
Using mailbox area	2·85*	3·56	3·61***
1-1 meetings with supervisor	2·36	2·70**	3·21***
Large group meetings	2·11	2·25**	3·29***
Small group meetings	2·68	2·68**	3·79***
Working on project with other	2·54*	3·04**	3·61***
Answering others' phones	2·18*	3·36	3·43***
3. Percentages of people reporting additional space needs:			
More individual work space needed	54%	21%	4%
More worktop surface needed	75%	18%	14%
More private (unit) meeting spaces needed	57%	43%	11%
More space to work together on projects needed	50%	32%	11%

* Significant difference between Phase I and Phase II.
** Significant difference between Phase II and Phase III.
*** Significant difference between Phase I and Phase III.

Table 3.3 Satisfaction with access.

	Phase I	Phase II	Phase III
Satisfaction with access to people, equipment and resources (5-point scale):			
Access to materials/equipment	2·54	3·00	3·07***
Access to people	2·82*	3·71	4·00***
Effect of my location in feeling like part of the team	2·71*	3·29	3·50***
Ease in getting from my workspace to other areas of the unit	2·86*	3·71	3·86***
Ease in performing access-related activities (4-point scale):			
Getting in and out of unit	2·96*	3.50	3·71***
Getting through unit aisles	2·39	2·54**	3·39***
Getting to mail	2·93*	3·54	3·57***
Getting to supplies	2·36*	2·96	3·17***
Getting to VDT	2·96*	3·46	3·46

* Significant difference between Phase I and Phase II.
** Significant difference between Phase II and Phase III.
*** Significant difference between Phase I and Phase III.

Table 3.4a Satisfaction with current workstation (5-point scale).

	Phase I	Phase II	Phase III
Space			
Enough space to do job comfortably	2·50*	4·04	3·79***
Furniture			
Comfortable furniture	1·93*	3·93	4·04***
Lighting			
Comfortable lighting	2·50*	3·67	4·11***
Lighting adequate for VDT work	2·68*	3·29**	3·75***
Lighting adequate for microfiche work	2·29*	3·18	3·43***
Lighting adequate for general reading	2·93*	3·93	3·75***
Lighting adequate for reading small print or printouts	2·44*	3·58	3·57***
Satisfaction with glare-free workspace	1·71*	2·21	1·93
Satisfaction with shadow-free workspace	1·96*	2·52	2·46
Temperature and air quality			
Comfortable air temperature and quality	2·07*	1·43**	2·32

* Significant difference between Phase I and Phase II.
** Significant difference between Phase II and Phase III.
*** Significant difference between Phase I and Phase III.

Table 3.4b Employees reporting specific types of furniture and HVAC discomfort (%).

	Phase I	Phase II	Phase III
Furniture			
Chair back	68	29	21
Chair seat	64	14	11
Chair height	39	4	4
Worksurface space	50	7	7
Worksurface height	21	0	0
HVAC			
Too hot	46	75	54
Too cold	54	36	50
Too stuffy	75	82	57
Too drafty	21	4	32
Too smokey	46	46	0

Table 3.5 Satisfaction with privacy-related issues (5-point scale).

	Phase I	Phase II	Phase III
1. Control over access:			
Distraction from people stopping to chat	2·82	3·21	3·39***
2. Freedom from distraction:			
Satisfaction with:			
Ability to work without distractions	1·89*	3·19	3·43***
Ability to concentrate in unit	0·96*	1·68	1·89***
Ability to concentrate in individual space	1·07*	1·71	1·75***
Distraction from:			
Others' conversations	1·61*	2·32	2·50***
Phone conversations	2·57	2·93	3·21***
Unit traffic	2·74*	3·56	3·68***
Traffic close to my workspace	2·25*	3·19**	3·57***
VDT noise	3·44	3·57**	3·93***
Ringing phones	2·64	2·93	3·29***
3. Speech privacy:			
Satisfaction with:			
Speech privacy at other's desk	1·96*	2·79	2·86***
Speech privacy at my desk	2·18	2·54	2·61
Phone privacy in unit	1·86*	2·82	2·64***
Distraction from being able to be heard	2·29*	3·14	3·29***
4. Overall privacy in individual workspace	1·79	2·15	2·29***

* Significant difference between Phase I and Phase II.
** Significant difference between Phase II and Phase III.
*** Significant difference between Phase I and Phase III.

Table 3.6 Perceived effect of environment on productivity (5-point scale).

	Phase I	Phase II	Phase III
Specific effects on productivity			
Amount of work accomplished	2·75*	4·04	3·71***
Quality of work	3·04*	4·00	3·61***
Meeting deadlines	2·93*	3·93	3·64***
Frequency of errors	2·71	2·96**	3·36***
Taking responsibility	3·04*	3·57	3·36
Being creative	2·71*	3·64	3·57***
Overall positive effect on productivity			
Unit environment	1·96*	3·54	3·61***
Individual workspace	2·21*	3·79	3·93***

* Significant difference between Phase I and Phase II.
** Significant difference between Phase II and Phase III.
*** Significant difference between Phase I and Phase III.

Table 3.7 Productivity effects.

Phase (weeks)	Production	Hours	Productivity ratio	Productivity increase over Phase I (%)	Productivity decrease from Phase II (%)
I (52)	78 756	74 372·53	1·06		
II (81)	177 254	100 174·73	1·77	66·90	
III (19)	34 245	21 105·50	1·62	52·80	14·10
II and III (100)	211 499	121 280·23	1·74	64·20	

Almost 62% of the employees felt that the arrangement of their personal workspace had a negative effect on their productivity, while 60% felt that unit layout had a negative effect on the unit. Furniture discomfort was reported by 70% of this unit, while temperature and air quality were uncomfortable for 83%. Lighting and glare were also problematic for 55% (primarily when working with VDTs), while 75% of the unit reported being bothered by acoustical distractions and lack of speech privacy. When asked if, overall, the unit and personal work environment helped them do their jobs efficiently, negative responses came from 67% and 58%, respectively. Employees specified that the environment had a negative effect on (1) amount of work accomplished, (2) meeting deadlines, (3) frequency of errors, and (4) being creative. Appearance of the personal and unit space was also poorly rated.

Phase II — New furniture

The results of this study clearly show that when these workers changed from traditional furniture in bullpen spaces to workstations, ergonomic chairs and task lights, and a new layout (with easier VDT access), significant improvements occurred in their productivity as well as in their perceptions of their productivity, and in their satisfaction with the work environment.

After these Phase II environmental changes, productivity increased almost 67%. At about the same time as the environmental change, a reorganization occurred affecting the way this unit and several other adjacent units carried out their work, making it difficult to isolate the relative impact of facility changes on productivity. However, over two-thirds of unit employees said in interviews that the reorganization and the new furniture were equally responsible for their increased productivity. Unit management was more conservative in their estimates. They felt that new furniture was second in importance, after the reorganization, and probably resulted in a 10–15% productivity gain. Some employees and supervisors said that had the organizational change occurred without environmental supports such as increased personal space and privacy, productivity would have decreases. These perceptions were confirmed by supervisors in adjacent units which experienced the same reorganization changes but no environmental changes. Some supervisors reported that although productivity increased in their units, it would have increased more with the help of environmental improvements. Other supervisors reported productivity decreased after this reorganization.

Employee satisfaction with the work environment also improved significantly at Phase II. Survey results show that they perceived the space as much more adequate, functional and professional-looking. Employees also reported that the workstations and new layout helped them work more efficiently and made access to people and resources easier. Satisfaction with comfort increased most dramatically during Phase II in the areas of (1) space, (2) furniture, and (3) lighting. As in the Renovation POE, reactions to task lights were mixed. No one was very comfortable using unadjustable side chairs to work on VDTs in multiple use areas. The increased privacy and freedom from distraction, provided by workstation panels, was one outstanding benefit of Phase II. In interviews employees discussed their increased satisfaction with privacy as follows:

> 'Panels create a sense of having your own office, which makes you want to do a better job.'
>
> 'People work better when allowed some privacy.'
>
> 'Increased privacy allows you to concentrate. You can accomplish more and do more accurate work.'

The new workstations and layout also increased employee pride in their individual and unit space, and in their satisfaction with it. Employees commented:

> 'It finally feels like we're being taken seriously.'
>
> 'I now feel that I have more impact.'
>
> 'New furniture makes you feel more professional and increases your morale by providing a sense of pride in the workplace.'

In the survey, employees also specified that Phase II changes had a more positive effect on the amount and quality of their work, on meeting deadlines, on taking responsibility and on being creative.

Phase III — Renovated space

Unit productivity levelled off somewhat after the move to totally renovated space, but was still 53% higher than it was in old space. Interviews with supervisors suggest that productivity decreases were due to two factors. (1) Unit output at Phase II was somewhat elevated by an effort to reduce a large backlog. This was accomplished by Phase III, and had some effect on reduced output. (2) Early in Phase III, a second and more encompassing reorganization was announced. Employee uncertainty about the implications of this caused some decrease in morale and in productivity. Productivity in adjacent units also declined after this announcement.

The fact that overall Phase II and Phase III unit productivity averaged 64·2% above Phase I during periods of reorganization and insecurity may suggest a stronger relationship between environmental improvements and increased productivity. Employee interview comments support this relationship.

> 'It's the best thing that ever happened here.'

'. . . has had a positive effect on the volume, speed and accuracy of my work.'

'. . . has given the whole unit a lift — an absolutely positive effect on performance.'

'. . . increases morale by providing a sense of pride in the workplace.'

'. . . tends to make you more professional.'

'. . . increases incentive.'

'. . . makes you want to come to work.'

During Phase III, further significant improvements occurred in employee environmental satisfaction and perceptions about their productivity. During Phases I and II, the study unit had occupied over 7500 ft^2 of space. When they moved into Phase III renovated space, they were able to be accommodated in 5600 ft^2. Despite the loss of almost 2000 ft^2 of unit floor space, employee satisfaction with unit space adequacy rose even higher than at Phase II. Space for meetings was also much more satisfactory at Phase III, with the introduction of oases. Employees and supervisors pointed out, however, that oases were not appropriate for meetings requiring privacy.

Traffic patterns were significantly improved during Phase III, as VDTs moved out of aisles and into workstations. Indirect overhead lighting increased satisfaction with lighting for VDT work, but glare filtering window shades did not improve ratings on glare. Educating employees on the importance of keeping shades drawn might improve this situation. Denser shades may also be needed in sun-intensive areas. As in the POE, reactions to keyboard arms were mixed; these have now become an optional feature. Temperature and air quality comfort did not improve as much as expected. The heating, ventilation and air conditioning system will be re-evaluated when the renovation is completed.

Distractions from traffic close to the personal workspace decreased at Phase III, with sound-masking and the move of VDTs out of the aisles. The placement of VDTs in workstations (to be shared with one other person by way of carousel platforms) was more desirable to these employees than having them in multiple use areas for several reasons. There was more privacy and less distraction in the workstations and employees were more comfortable using their ergonomic desk chair, rather than their side chair, for VDT work. Employees commented on their surveys and in interviews that having a 1:1 terminal-to-worker ratio would be ideal for concentration, productivity and, ultimately, customer service.

Achieving the right balance between privacy and interaction can be critical to productivity. Throughout the study, satisfaction improved significantly in both these areas. However, even by Phase III, ratings on (1) speech privacy, (2) ability to concentrate, and (3) the overall privacy provided by individual workspaces did not reach the midpoint on the five-point scale of satisfaction. Possible explanations include: (1) sound-masking may not be as effective as was originally hoped, and (2) the types of spaces provided by the renovation (common footprint workstations with three-sided enclosure and oases) may not offer enough privacy to some employees. Although working styles at Aetna appear to be moving more and more in the direction of project team interaction, these results suggest that more options for achieving privacy as well as for team interaction should be included in future environmental planning.

Productivity results

Table 3.7 shows (1) the total amount of production (number of forms processed) and cumulative work hours for each of the three phases, and also for Phases II and III combined, (2) the average ratio of production to hours for each period (productivity ratio), (3) the percent increases in productivity (above Phase I) for Phase II, and Phase III, and for Phases II and III combined, and (4) the decrease in productivity from Phase II to Phase III.

As seen in Table 3.7, at Phase II, productivity increased 67% above the value for Phase I. Although productivity slipped at Phase III, it was still 53% above its Phase I level. Supervisors attribute this slip to a levelling off of the work backlog and to employee insecurity about the second reorganization. When Phases II and III are taken together, they show a productivity increase of 64% above Phase I. It is obvious that this increase is substantial.

Effects on absenteeism

Absenteeism was studied in order to accurately estimate hours worked for the productivity analysis, and also to discover whether it changed when the environment was improved. Due to the difficulty in obtaining detailed absence records for people who left the unit while the study was in progress, we did not differentiate between absenteeism due to vacations and occasional illness and prolonged absenteeism which might suggest job dissatisfaction. Comparing absenteeism ratios across the three phases (total hours absent over total possible work force hours) we noted that absenteeism decreased 14% from Phase I to Phase II. Since employees commented that improved space increased their sense of professionalism and made them want to come to work, it is possible that Phase II improvements played some role in decreased absenteeism. During Phase III, however, absenteeism rose 7% above its Phase I level. Phase III encompassed the last 4·5 months of 1988, and thus is less comparable to the other phases. When we compare the last 4·5 months of 1988 with the same periods in 1987 (Phase II) and 1986 (Phase I), there is very little difference. For this unit, absenteeism seemed to be higher during the last 4·5 months of the year, than at other times of the year.

Effects on turnover

Some amount of turnover is healthy in any organization. Generally, a steady turnover rate is associated with normal job satisfaction, while increased turnover is associated with dissatisfaction. Although our results show a 21% increase in turnover in Phase II and an 86% decrease in Phase III, the highest turnover occurred at the very end of Phase I and during the first 6 months of Phase II. Unit management attributes this increased turnover to the initial reorganization.

Cost–benefit analysis

The Phase II furniture costs were approximately \$5000 per employee, or \$195 000 for the 39 member unit. Phase III renovation costs were \$9100 per employee or \$327 600 for the 36 member unit. Since we could not isolate the relative impact of facility change versus organizational change on productivity, we made assumptions in order to examine renovation costs against productivity benefits.

Employees attributed half their productivity increase to environmental change. Given average Phase II salaries, half of the Phase II and Phase III productivity increases are worth \$621 100 compared to the \$522 600 cost of environmental changes. Thus, under the employees' assumptions, the renovation costs are paid for by half of the productivity increase, with a net benefit to the company of \$98 500 during Phase III.

Management attributed a 10–15% productivity increase to environmental change. Under this assumption, all environmental improvements are paid for by 1992, with a net benefit to the company of \$110 000.

Discussion

Strengths and limitations

Our methods of examining and analysing environmental satisfaction and productivity were systematic. The attitudinal survey was the product of a literature review, employee interviews, behavioural observations and pilot testing. The participants and their work were representative of the insurance industry. Many aspects of environmental satisfaction and productivity were examined and reported.

The fact that this study was conducted in a real-life Aetna environment, as opposed to a controlled laboratory setting, is both a strength and a limitation. Typical of most Aetna groups, this department changed during this 3-year study in other ways besides environmentally. Some of these changes occurred gradually throughout the study in the areas of personnel, equipment and software. A more dramatic change occurred at the beginning of Phase II in the form of a reorganization in the way the unit carried out its work. These other changes probably had little or no effect on environmental satisfaction or on personal reports about the influence of the physical environment on productivity (as measured in the attitude survey). However, they probably did influence actual productivity. This made it difficult to estimate the exact effect of the environmental changes versus that of other changes on productivity.

Nevertheless, the fact that significant changes in productivity occurred in a real-life setting meant that the results were more generalizable to other Aetna settings where organizational change has frequently been the mode of existence. Although we did not have data from a control group to compare with our results, interviews with supervisors in other units, which had experienced all other changes except environmental, suggest that at least part of our unit's productivity increase was caused by environmental changes.

Our understanding and interpretation of the effects of environmental and organizational change on productivity, absenteeism and turnover has been enriched by the responses to survey questions that directly addressed the effect of the environment on productivity, as well as through periodic interviews with unit employees, and discussions with representatives of adjacent units that experienced organizational changes but not environmental changes.

Implications

Aetna employees are indicating through surveys and interviews that they are experiencing significant increases in comfort, environmental satisfaction and productivity as a result of improvements to their workspace. The Productivity Study results suggest a strong relationship between environmental improvements and productivity increases. This study also clearly shows that environmental improvements caused significant increases in employee satisfaction with that environment.

When the unit under study moved from traditional furniture in bullpen space to workstations, their ratings of many environmental features rose significantly. Such improved ratings occurred in the areas of:

(a) individual and unit space adequacy (including functionality),
(b) access to people and resources,
(c) comfort from space, furniture and lighting, and
(d) acoustical and visual privacy.

In addition, these employees reported that improved surroundings had a significantly more positive effect than the old space on:

(a) amount and quality of work,
(b) meeting deadlines,
(c) taking responsibility, and
(d) being creative.

When the unit moved into totally renovated space, ratings improved significantly in the areas of:

(a) unit space adequacy,
(b) adequacy of meeting space,
(c) traffic circulation through unit aisles,
(d) lighting comfort for VDT work, and
(e) distractions.

Employees also reported that in totally renovated space, errors were less frequent.

Despite our efforts, some problems still exist in areas such as HVAC discomfort, acoustical privacy and distraction and glare. Most of these problems are being further researched and solutions will be evaluated.

Aetna's research efforts have recently been enriched by the addition of a permanent 'laboratory' space in which to test both environmental supports and computer hardware and software. This laboratory is a joint effort of Facilities Planning and Corporate Technology Services.

Other areas currently being examined at Aetna include the adequacy of the common footprint workstation for people with varying types of responsibilities, the effectiveness of alternative workspaces such as the oases, and the education of employees on how to be most comfortable and productive in renovated space.

References

Altman, I., 1974, Privacy: a conceptual analysis, In: *EDRA 5: Man–Enviroment Interactions: Evaluations and Applications, The State of the Art in Environmental Design Research — 1974*, edited by D. H. Carson, pp. 3–28 (Milwaukee: Environmental Design Research Association).

Becker, F. D., 1981, *Workspace: Creating Environments in Organizations* (New York: Praeger).

Becker, F. D., 1986, Work in its physical context: the politics of space and time. In: *The Impact of the Work Environment on Productivity*, edited by M. E. Dolden and R. Ward, Jr, pp. 23–49 (Washington, DC: Architectural Research Centers Consortium, Inc.)

Brill, M., Margulis, S. T. and Konar, E., 1984, *Using Office Design to Increase Productivity* (Buffalo, NY: Workplace Design and Productivity)

Hedge, A., 1982, The open-plan office: a systematic investigation of employee reactions to their work environment. *Environment and Behavior*, **14** (5), 519–542.

Justa, F. C. and Golan, M. B., 1977, Office design: is privacy still a problem? *Journal of Architectural Research*, **6** (2), 5–12.

Laufer, R. S., Proshansky, H. M. and Wolfe, M., 1973, Some analytic dimensions of privacy. *Proceedings of the 3rd Annual Architectural Research Conference*, Lund, Sweden.

Louis Harris and Associates, Inc., 1978, *The Steelcase National Study of Office Environments: Do They Work*? (Grand Rapids, MI: Steelcase).

Louis Harris and Associates, Inc., 1980, *The Steelcase National Study of Office Environments No. II: Comfort and Productivity in the Office of the 80's* (Grand Rapids, MI: Steelcase).

Proshansky, H. M., Ittelson, W. H. and Rivlin, L. G., 1970, Freedom of choice and behavior in a physical setting. In: *Environmental Psychology*, edited by H. M. Proshansky *et al.* (New York: Holt, Rinehart and Winston).

Sundstrom, E., 1986, *Work Places* (New York: Cambridge University Press).

Westin, A., 1970, *Privacy and Freedom* (New York: Atheneum).

Wineman, J. D., 1986, New developments in office setting research. In: *The Impact of the Work Environment on Productivity*, edited by M. E. Dolden and R. Ward, Jr, pp. 71–92 (Washington, DC: Architectural Research Centers Consortium, Inc.).

Chapter 4

Ergonomic Improvements in VDT Workstations: Health and Performance Effects[1]

MARVIN J. DAINOFF

Introduction

The decade of the 1970s may well be remembered as the period of initial transition into the computerized office. It was during this time that automated information processing procedures (e.g. airline reservations, word processing, account maintenance) began to be implemented on a relatively large scale, and to involve non-computer professionals in daily interaction with computer equipment via some sort of video display terminal (VDT). However, as the spread of VDTs into the workplace escalated, so did associated health complaints by the new operators, to the point where, even in 1977, trade magazines could write about 'epidemics' of such complaints (see Dainoff and Dainoff, 1986).

By 1981, there was a relatively extensive literature indicating the presence of some attribute or set of attributes characteristic of work at VDTs which impels those performing such work to complain rather frequently of a variety of visual and musculoskeletal disorders. Dainoff (1982) reviewed much of this literature.

Having made this statement, it remains to identify what these attributes are. In point of fact, the existing literature (prior to commencement of this research in 1980) had not been terribly helpful in providing direct causal linkages between specific aspects of VDT workplace design and specific complaints/symptoms. Several studies (e.g. Coe *et al.*, 1980; Grandjean, 1980; Smith *et al.*, 1981) had shown higher visual and musculoskeletal complaint levels among VDT workers than among comparable workers not using VDTs. On the other hand, these studies have been criticized (e.g. National Research Council, 1983) as to adequacy of control (are the non-VDT workers truly comparable?) and other more recent studies have failed to find differences in situations where VDT operators are matched against controls (Starr *et al.*, 1982; Gould and Grischkowsky, 1983).

[1] This research was conducted while the author was a Visiting Researcher in the Motivation and Stress Research Station, Applied Psychology and Ergonomics Branch, Division of Biomedical and Behavioral Sciences, National Institute for Occupational Safety and Health. The research was part of a programmatic effort by the Institute towards the goal of stress reduction among VDT operators.

The position taken here is that posing the argument in terms of a comparison between VDT and non-VDT operators does not seem to be particularly useful. Introducing a VDT-based information processing system into a workplace will inevitably generate a host of changes in the physical, psychological and social structure of that work. Therefore, discussions over comparability of control groups are, in field studies, inherently difficult to resolve since we can rarely be clear on exactly (or even approximately) what we are comparing. This difficulty extends to the laboratory. For example, if one is setting up a laboratory comparison of VDT/non-VDT text editing, should not the control group be looking at copy which is vertically oriented and covered with a glass plate?

However, if we look closely at both sides of the argument, we can find some very important information. Those studies finding differences between VDT and non-VDT work have also tended to characterize the VDT workplaces included in their investigations as being deficient with regard to ergonomic attributes. On the other hand, each of the studies described above which *failed* to find a difference between VDT and non-VDT operators also indicated that ergonomic conditions were relatively good among the former. Therefore, it would seem that the highest research priority would be not to generate VDT/non-VDT comparisons, but rather to attempt to clearly isolate those ergonomic factors which differentiate good from poor VDT workstations, and to determine if such differences are linked to operator health problems (see Dainoff, 1984, for a more extended version of this argument). This logic forms the basis for the current research.

Rationale

Ergonomics can be conceptualized as the *fit* between the person and his/her physical environment (Dainoff and Dainoff, 1986). As such, this environment must be considered ecologically as a system of interconnected components. It is assumed that, when the system is optimal, operator efficiency is also optimal in the sense that work performance is high and physical effort is (relatively) low. As the system is degraded, the resulting effect is some combination of increased physical strain, perceived fatigue and decreased performance (Grandjean, 1979).

Within this framework, the current research represents an attempt to converge on specific differences between ergonomically optimum and suboptimum VDT workplaces. Given that there is some measure of agreement with respect to the importance of lighting, glare control and worksurface/chair adjustability in providing an optimum workplace, the goal of this research is to set up an experimental comparison in which all factors remain constant except those specific ergonomic factors. We created, therefore, two workstations. One was ergonomically optimal with respect to the above factors; the second suboptimal. A group of subjects then alternates between these workstations, performing the same date-entry task in each. The working sessions are sufficiently long (3 h/day over a 5-day week) to provide at least a reasonable laboratory approximation to actual working conditions.

We then ask a simple, but global question: Does this cluster of ergonomic factors differentiating these workstations have any measurable effect on the operators who work in them? Given that the initial global question has a positive answer; that we have, in fact, found that good ergonomic design can make a difference, we can converge on specific classes of ergonomic attributes (i.e. we hold lighting constant and vary only adjustability of worksurfaces). The following two experiments represent an implementation of this approach.

Methodology

General design

The basic experimental design consisted of a global comparison between two VDT workstations which differed in their ergonomic attributes. At one workstation, these attributes could be characterized as 'optimal'; at the second, as 'suboptimal'. In Experiment 1, the optimal–suboptimal difference involved both furniture and glare control; in Experiment 2, only furniture was involved.

A repeated measures approach was utilized in which each subject worked at a data-entry/editing task for two 3-h sessions at each workstation in counterbalanced order. A 5-day week (Monday to Friday) was required to run each pair of subjects. The first day (Monday) served as a training/baseline day, while, for the remaining days, the subjects alternated between optimal (O) and suboptimal (S) workstations. Table 4.1 outlines the overall design.

Table 4.1. Design Summary: weekly schedule for two subjects. (O) *indicates optimal workstation;* (S) *indicates suboptimal workstation.*

Subject	Monday	Tuesday	Wednesday	Thursday	Friday
1st	Training	O	S	S	O
2nd	Training	S	O	O	S

Before and after each 3-h work session in Experiment 1, there was a 30-min period allotted to a battery of tests designed to assess visual function (visual contrast sensitivity, flicker, near points of convergence, and accommodation, lateral phoria, acuity), blood pressure (systolic and diastolic) and subjective discomfort (visual, musculoskeletal, headache). Finally, at the conclusion of the final (Friday) post-experimental assessment battery, a short debriefing interview was conducted. Since the visual function and blood pressure assessment data yielded no differences between workstation conditions, these procedures were not carried out during Experiment 2. Results from these procedures will not be discussed in this chapter.

General procedure

At the beginning of each work day, subjects either were administered the visual function/blood pressure tests (Experiment 1) or proceeded directly to their workstations. Subjects then filled out a 10-item discomfort questionnaire which appeared on their VDT display. Three of the items related to vision complaints, six of the items related to musculoskeletal complaints, and the final item related to headache.

Upon completion of the discomfort questionnaire, the subjects began a 3-h automatically-timed data-entry session. The data-entry task, which simulated a chemical stock inventory, consisted of two subtasks: file entry and file maintenance.

The file-entry subtask required the subject to enter, from paper copy, elements of information onto a 'stock entry' form seen on the display screen. Each screen consisted of 10 data fields, the information to be entered therein included the name of the chemical, a 10-digit alphanumeric stock number, a 10-digit location code, a 4-digit bin number, a price and the name, address and phone number of the manufacturer. Information for each individual display screen was contained on a separate sheet of paper copy, which was turned over after all data fields were filled in.

During the file maintenance subtask, the informational structure of the screen display and paper copy was the same as for file entry. However, in this task, the display screen was already filled in and the subject was required to carry out a field by field comparison between screen and paper copy to verify accuracy of the information on the screen. In fact, errors were actually present in seven of the ten data fields on each maintenance file. Error location was random with the constraint that the likelihood of an error appearing in a given field was approximately constant across the entire set of maintenance files.

During the 3-h session, subjects worked on alternating sets of 20 entry files and 20 maintenance files without a break. At the conclusion of 3 h, they again filled out the discomfort questionnaire, and, for those in Experiment 1, completed the post-experimental vision/blood pressure assessment procedure.

Incentive pay

The subjects were all operating under incentive pay conditions. They were aware that baseline data, in keystrokes per minute, had been collected during the training session on day 0, and that they would receive incentive pay as hourly increments to their base salary for a given day as follows:

$$I = (0 \cdot 10)KPM - (0 \cdot 4)EPM$$

where I = incentive pay per hour (dollars), KPM = keystrokes per minute over baseline rate, and EPM = errors per minute.

One or more mistakes (missing or changed values) in any of 10 data fields was counted as a single error. Subjects were informed of their incentive pay increments midway through each session and again at the end of each session.

Training

Monday, as the first day of the experimental week, was devoted to training and baseline performance measurement for each pair of subjects. In Experiment 1, training began with an orientation to the visual function and blood pressure assessment procedures. This was followed by an introduction to the workstation, the discomfort checklist, and the task. Experiment 2 subjects went directly to this stage of training. After a very detailed review of procedures for both entry and maintenance subtasks, subjects were asked to practise both subtasks for about 2 h; alternating between subtasks as they would later in the actual experimental session. They were instructed to '. . . type as fast as you can but be as accurate as you can'. At no time, up to this point, was any mention made of incentive pay, baseline conditions, ergonomic/adjustable furniture or lighting conditions. Subjects' performance levels during these 2 h became the baseline data from which incentive pay increments would be computed during the actual experimental sessions.

In order to insure that ergonomic conditions were similar for both subjects during the training/practice phase, each workstation was 'normalized' in the sense that lighting conditions were made equivalent, seat pan heights were both set to 47 cm, and keyboard and monitor support surfaces were adjusted to 72 cm.

At the conclusion of the practice sessions, both subjects were then informed of the incentive pay system, and given careful instruction as to its operation. In addition, the subject who was to start the experimental sequence in the optimal condition, was shown how to adjust the chair for optimum posture. The experimenter, in turn, adjusted the worksurface levels so as to achieve the recommended 90° optimum posture. (This ergonomic instruction/adjustment was delayed for the second subject until the beginning of Wednesday's session, when it would be her/his turn in the optimal workstation.)

Apparatus

Workstation characteristics — Experiment 1

Equipment of ergonomic relevance used in setting up workstations included an IBM Synergetics adjustable VDT terminal base and task chair, a Wright Line ETS 2000 adjustable terminal base, inline copy holder and wrist rest, an Armstrong Tascam parabolic lighting fixture, a Polaroid CP70 contrast enhancement filter, an Acme-Lite music stand light, and two DEC VT100 VDTs.

Each workstation was set up in an enclosed room measuring 229 × 196 cm for the optimal and 236 × 195 cm for the suboptimal. Each room contained an adjustable terminal base, chair, overhead lighting fixture and a TV camera.

OPTIMAL WORKSTATION

The parabolic lighting fixture was located at a height of 221 cm within the optimal workstation. The music stand light, used to provide additional illumination on the paper copy, was attached to the top of the inline copy holder located between the keyboard and screen. The task chair utilized was gas-cylinder adjustable in both seat pan height and back tilt, with adjustable lumbar support pad located 21 cm from seat top. Total backrest height was 33 cm. A wrist rest was located directly in front of the keyboard. Chair, keyboard height, and screen height were individually adjusted so that subjects' feet were flat on the floor, their lumbar region of the back was supported, their elbow height was approximately at the level of the home row of the keyboard, and their viewing angle (eye to screen centre) was approximately 20° from the horizontal. The contrast enhancement filter was placed over the face of the terminal screen, and the VDT contrast control increased such that a character–screen contrast ratio of approximately 8 : 1 was obtained. Illumination on the paper copy was approximately 550 lx. (Luminance and illuminance were measured with a Spectra Photomate Spotmeter.)

SUBOPTIMAL WORKSTATION

A standard issue GSA tilt and swivel office chair was utilized. Adjustment of seat pan height could not be accomplished from a seated position; it was, therefore, maintained at 47 cm throughout the study. The backrest, which was 23 cm high and 9 cm above the seat pan, did not contain a lumbar support, and was not adjustable. In addition, there was a considerable degree of 'give' in the back rest; this was in contrast to the positive support provided by the optimal chair. Copy holder and wrist rest were not utilized. The adjustable keyboard and monitor support surfaces were deliberately maladjusted such that forearm angles were approximately 45° and screen viewing angle approximately 30° from the horizontal. Glare conditions were created by removing the diffuser from a four-tube bank of overhead fluorescent lights (at a height of 232 cm), and lowering the contrast control on the VDT. This resulted in a character–screen contrast ratio of slightly better than 1:1, with an illuminance of 1081 lx on the paper copy. (It should be emphasized that these 'suboptimal' conditions were quite similar to many actual workstations observed in the field at the time of the study.)

Workstation characteristics — Experiment 2

Lighting attributes in both workstations were changed from their Experiment 1 configurations so that glare conditions were similar. To accomplish this, the glare was reduced in the suboptimal environment by removing one tube from the four-tube bank of overhead fluorescent lights (height, 232 cm) and replacing the diffuser which had been removed during Experiment 1. A compensatory correction was required in the optimal environment; this was accomplished by removing the contrast enhancement filter from the VDT screen, as well as adjusting the contrast control. The parabolic lighting fixture (height, 221 cm) remained unchanged. The resulting illuminance values on the paper copy were 56 ft-candles in the optimal workstation and 59 ft-candles in the suboptimal. Corresponding character to background contrast ratios were 1·33 and 1·22. Luminance and illuminance were measured with a Spectra Photomate Spotmeter.

OPERATING SYSTEMS AND RELATED EQUIPMENT

Task control, performance monitoring, and discomfort scale recording was accomplished via an operating system consisting of three Apple II Plus microprocessors in conjunction with a Corvus Constellation 5·5 megabyte hard disk/multiplexer. This system was used to control the Digital Equipment Company VT100 VDTs located in the two workstations. A Panasonic closed circuit video tape recording system was employed to allow continuous visual monitoring of each subject.

Subjects

General characteristics

Trained (40–55 words/min) typists with no prior VDT experience were hired from an employment agency specializing in temporary clerical help. They were informed, by their employer, that they would be participating in a study related to occupational stress in clerical work; however, the terms ‘ergonomic’ or ‘workplace design’ were not used. Pre-selection physical and ophthalmological examinations were conducted; insuring that all subjects were in general good health with no histories of hypertension and/or musculoskeletal disorders, that corrected visual acuity was at least 20 : 20, and that no evidence of presbyopia requiring the use of bifocal lenses was seen. The last criterion was utilized so that the study might not be confounded by the special postural problems often seen in VDT work by wearers of bifocals.

Experiment 1

Twelve subjects were female; one was male. Median age was 26, with a range of 19–35 years; median height was 162·6 cm, with a range of 144·8–171·8 cm, and median weight was 61·9 kg with a range of 49·1–74·5 kg. These may be compared with recent US civilian female median population heights of 160·5 cm, and weights of 61·1 kg (Kroemer and Price, 1982); indicating that the present sample is not anthropometrically unrepresentative. The male subject's height was 162·6 cm and weight was 58·5 kg.

Experiment 2

Fourteen of the subjects were female, and 1 was male. Median age was 32, with a range of 18–51 years; median height was 163·84 cm, with a range of 156·16–188·16 cm; and median weight was 57·61 kg with a range of 46·04– 92·99 kg. As above, comparing these values with the Kroemer and Price (1982) tables indicates that the present sample is not anthropometrically unrepresentative. The male subject's height was 188·16 cm; his weight was 92.99 kg.

Final interview

At the conclusion of the post-experimental testing on the last day, a final interview and debriefing was conducted with each subject. The interview consisted of a non-directive probing of each subject's impressions of the study and the workstation. Each subject was asked: 'What did you think was going on in this study?' If her/his response did not refer to ergonomic differences between workstations, she/he was asked: 'What did you think about the two rooms you worked in?'

Results

Because of cost and administrative concerns, the number of subjects we were able to employ in each study was necessarily limited. Moreover, the dependent measures employed — performance, and subjective complaints — differed with respect to their relative statistical power. Given this concern, together with the exploratory nature of the study, and the relative practical costs associated with Type I and II error, the decision was made to adopt a loose ($P = 0 \cdot 10$) level of statistical significance for the questionnaire data.

In addition, some of the effects were seen to differ depending on whether they occurred during the first 2 days of the study or the second 2 days. The counterbalanced design structure allows each half of the 4-day sequence to be treated as a separate replicate; first or second half will, therefore, be treated as a classificatory variable in all analyses.

Performance

General comments

While the decision to employ incentive pay was primarily based on the need to maintain subject motivation, the manner in which incentive pay was computed (keystroke rate less error rate) made it a reasonable primary dependent measure. In fact, a standard text (McCormick and Ilgen, 1980, p. 50) prefers changes in performance rate as a general criterion measure.

Incentive pay figures were computed for each subject at the end of each of the four daily experimental sessions. As discussed earlier, incentive pay (I) consisted of two weighted components:

$$\mathrm{I} = (0 \cdot 10)\,\mathrm{KPM} - (0 \cdot 4)\,\mathrm{EPM}$$

The keystroke rate component (KPM) was obtained by dividing average number of keystrokes per file by average time per file. From this value was subtracted the keystroke rate obtained on the initial baseline day. The error rate component (EPM) was obtained by dividing average errors per file by average time per file. Each component was then inserted into the equation to obtain a weighted hourly increment in incentive pay.

For example, assume a given subject obtained a (typical) baseline performance of 50 keystrokes/min during the initial training session (day 0). On day 1, she finished the day with an average performance of 60 keystrokes/min, an average of 0·5 errors/file, and an average time of 1 min/file. The KPM component would consist of 10 keystrokes/min (over baseline) weighted by US$0·10, yielding a value of $1·00.

Table 4.2 Experiment 1. Average incentive pay (per hour) within ergonomic condition (optimal, suboptimal); by session half and combined across days. In combined data, percent O/S defined by ((suboptimal–optimal)/optimal) × 100.

	First half		Second half		Combined			
Subject	Suboptimal	Optimal	Suboptimal	Optimal	Suboptimal	Optimal	Difference	%O/S
3	3·53	2·20	4·26	5·07	3·90	3·64	−0·26	−6·68
4	1·30	2·77	2·99	3·06	2·15	2·92	0·77	35·90
5	2·67	1·99	3·47	4·28	3·07	3·14	0·06	2·12
6	2·61	1·91	3·10	3·91	2·86	2·91	0·06	1·93
7	1·42	2·00	2·78	2·28	2·10	2·14	0·04	1·90
8	1·84	1·64	2·11	2·70	1·98	2·17	0·19	9·87
9	0·76	2·10	1·79	2·53	1·28	2·32	1·04	81·57
10	2·67	2·66	3·21	4·30	2·94	3·48	0·54	18·37
11	0·64	1·49	1·45	1·81	1·05	1·65	0·61	57·89
13	0·51	1·70	2·36	2·60	1·44	2·15	0·72	49·83
14	2·87	2·08	3·17	3·87	3·02	2·98	−0·04	−1·49
15	0·77	2·80	3·66	3·13	2·22	2·97	0·75	33·86
17	1·15	2·39	3·37	2·92	2·26	2·66	0·40	17·48
Mean	1·75	2·13	2·90	3·27	2·33	2·70	0·37	23·27

From this would be subtracted the EPM component (0·5/1·00 = 0·5) weighted by US$0·40, yielding $0·20. Thus, at the end of day 1, an I value of US$0·80 (1·00 – 0·20) was earned by the subject. This was added, as an hourly increment, to her pay for that day.

Experiment 1

Table 4.2 indicates average levels, and optimal–suboptimal differences in incentive pay for both experimental halves as well as for differences combined across days (e.g. days 1 and 4 vs. days 2 and 3). For all three analyses, performance in the optimal workstation is superior to performance under suboptimal conditions in that more incentive pay is earned. In the combined analysis, a correlated t-test indicated that the mean improvement of US$0·38 in incentive pay increment was statistically significant; $t(12) = 3{\cdot}44$, $P = 0{\cdot}005$.

An additional way of describing the degree of improvement was to compute the percentage of increase in incentive pay earned for each individual ((O – S)/S × 100). For the combined conditions, percentage of increased incentive pay as a result of working under optimal conditions ranged from slightly negative (– 2·6%) to a high of 81·1%. The average percentage improvement was 23·3%. In preliminary reports (e.g. Dainoff, 1983a,b), this figure is reported as 24·7%. The discrepancy is due to rounding error.

While incentive pay can be viewed as a global performance measure, a more fine-grained assessment was carried out on the components of incentive pay: keystrokes per minute and average errors per file. Independent repeated measures 2 × 2 analyses of variance were carried out on each of these components. Factors were session half (H; first/second) and ergonomic condition (C; optimal (O)/suboptimal (S)).

It appears that virtually all of the improvement in performance reflected in the global incentive pay measure was due to an increase in keystroke rate. Obtained F ratios for keystrokes per minute indicated that the condition factor was highly

Table 4.3. Experiment 1. Means (M) and Standard Errors of the Mean (SEM) for keystrokes per minute and errors per file by session half (first/second) and condition (optimal/suboptimal).

	Keystrokes		Errors	
	M	SEM	M	SEM
1st half				
Optimal	79·43	1·48	0·644	0·1
Suboptimal	75·58	3·42	0·681	0·1
Difference	3·85		–0·037	
2nd half				
Optimal	91·11	2·98	0·646	0·09
Suboptimal	87·30	2·24	0·626	0·07
Difference	3·81		0·02	

Table 4.4 *Experiment 2. Average incentive pay (per hour) within ergonomic condition (optimal/suboptimal); by session half and combined across days. In combined data, percent O/S defined by ((suboptimal–optimal)/optimal) × 100.*

	First half		Second half		Combined			
Subject	Suboptimal	Optimal	Suboptimal	Optimal	Suboptimal	Optimal	Difference	%O/S
14	0·83	2·37	2·73	3·18	1·78	2·78	1·00	55·90
15	2·36	1·69	2·83	4·42	2·60	3·06	0·46	17·73
16	1·56	1·90	2·75	2·97	2·16	2·44	0·28	12·99
17	1·63	2·83	3·99	3·57	2·81	3·20	0·39	13·88
18	2·04	1·62	1·85	2·66	1·95	2·14	0·20	10·03
19	1·45	3·05	4·11	4·10	2·78	3·58	0·79	28·60
20	0·72	0·31	1·45	1·98	1·09	1·15	0·06	5·53
22	2·16	0·94	2·56	2·84	2·36	1·89	−0·47	−19·92
23	1·11	2·28	3·44	3·08	2·28	2·68	0·41	17·80
24	1·08	0·81	1·53	1·98	1·31	1·40	0·09	6·90
25	1·05	1·84	2·73	2·14	1·89	1·99	0·10	5·29
26	2·23	1·85	2·60	3·50	2·42	2·68	0·26	10·77
27	0·71	2·37	2·48	2·69	1·60	2·53	0·94	58·62
28	1·89	1·09	2·33	3·26	2·11	2·18	0·06	3·08
29	0·59	1·52	1·84	1·78	1·22	1·65	0·44	35·80
Mean	1·43	1·76	2·61	2·94	2·02	2·35	0·33	17·53

significant: $F(1,12) = 10{\cdot}41$, $P = 0{\cdot}007$; as was the half factor: $F(1,12) = 102{\cdot}32$, $P = 0{\cdot}00001$. At the same time, the F ratio for the H × C interaction approached zero. However, in the case of average number of errors, all of the F ratios (main effects and interaction) were less than 1·0.

Means and standard errors for both of these measures can be seen in Table 4.3. It might be pointed out that the error data in Table 4.3 reflect the average number of errors per file. Since there were 10 fields per file, these values, multiplied by 10, are equivalent to average per cent error per condition. Combined average per cent error, across all conditions, was 6·49%.

Experiment 2

Table 4.4 indicates average levels, and O – S differences in incentive pay for both experimental halves as well as for differences combined across days (e.g. days 1 and 4 vs. days 2 and 3). As in Experiment 1, all three analyses indicate that working under optimal ergonomic conditions resulted in better performance (more incentive pay earned). In the combined analysis, a correlated t-test indicated that the mean improvement of US$0·45 in incentive pay increment was statistically significant; $t(14) = 3{\cdot}21$, $P = 0{\cdot}01$. For the combined analysis, the percentage of increase in incentive pay (O – S/S × 100) earned by each individual ranged from a low of – 19·9% to a high of 58·6%. The average percentage improvement was 17·6%; this may be compared with a comparable value of 22·3% found in Experiment 1.

Table 4.5 indicates keystrokes per minute and average errors by half and condition for Experiment 2. As in Experiment 1, independent repeated measures 2 × 2 analyses of variance were conducted on each of these dependent measures. It appeared that, once again, the bulk of the improvement in performance was due to increases in keystroke rate. Obtained F ratios were: for condition, $F(1,14) = 10{\cdot}93$, $P < 0{\cdot}005$; for half, $F(1,14) = 149{\cdot}29$, $P < 0{\cdot}00001$. The F ratio for the H × C interaction approached zero. Average number of errors decreased over the course of the

Table 4.5. Experiment 2. Means (M) and Standard Errors of the Mean (SEM) for keystrokes per minute and errors per file by session half (first/second) and condition (optimal/suboptimal).

	Keystrokes		Errors	
	M	SEM	M	SEM
1st half				
Optimal	74·72	4·66	0·78	0·09
Suboptimal	71·20	3·90	0·75	0·08
Difference	3·52		0·03	
2nd half				
Optimal	86·53	4·74	0·67	0·09
Suboptimal	83·17	4·44	0·69	0·08
Difference	3·36		– 0·02	

experiment — the main effect for half being significant; $F(1,14) = 6 \cdot 45$, $P < 0 \cdot 02$. However, neither the condition nor the H × C interaction for errors approached significance. Average combined per cent error, across all conditions, was 7·24%, which may be compared to the value of 6·49% seen in Experiment 1.

Discomfort questionnaire

Subjects responded to each of the 10 discomfort items displayed on their monitor screens by moving a cursor from left to right using the spacebar. The extreme left cursor position always reflected the least amount of discomfort. Consequently, the number of spacebar presses for each item was a direct measure of the amount of discomfort.

However, since our primary focus was on relative rather than absolute levels of discomfort, all analyses will be expressed as difference scores. Two levels of difference score were computed. First order differences consisted of post-session(after) responses less pre-session (before) responses. Second order differences consisted of suboptimal (before – after) differences less optimal (before – after) differences. On the assumption that subjects would be less comfortable after the session, and that this would be more likely in the suboptimal condition, the predicted signed direction for each difference is positive.

For example, assume that, prior to an experimental session under optimal conditions, the subject answers a given question by pressing the spacebar only two times. This would have positioned the display screen cursor very close to the 'not at all' end of the bipolar scale. At the end of the session, however, the key had been depressed 12 times, positioning the cursor more in the direction of the 'very much' end of the scale. The first order difference would be 10 for that question. If, for the same subject and question, the first order difference under suboptimal conditions, was 15, the corresponding second order difference would be 5.

Experiment 1

Table 4.6 contains the mean first and second order differences by condition and session half. Correlated one-tailed *t*-tests were run on all second order differences across subjects. Significant differences in the predicted direction were found for item 6 (pain in neck, shoulder and upper back), $t(12) = 3 \cdot 57$, $P < 0 \cdot 01$; and for item 8 (pain in lower back), $t(12) = 2 \cdot 11$, $P < 0 \cdot 01$, during the first half of the session. No significant differences were discovered during the second half.

Particular attention must be paid to the unexpected presence of negative first order differences in Table 4.6. Such differences reflect complaint levels which are greater at the beginning of the work session than at the end; consequently, interpretation of second order differences becomes problematic.

Examination of individual subject responses indicated that negative first order differences were much more frequent (9% vs. 2% incidence) for questions dealing with visual problems and headache (items 1–3,10) than for the remaining musculoskeletal questions (items 4–9). The most plausible explanation for the negative first order differences among the visual complaint items is that visual performance assessment procedures might well have been a source of short-term visual discomfort. These procedures, which included 15–20 min of determination of spatial frequency and flicker thresholds, generated several complaints among subjects and were, in fact, discontinued in Experiment 2. The fact that the pre-experimental visual performance assessment occurred *prior* to the discomfort checklist, but the post-experimental visual performance assessment occurred *after* the discomfort checklist, would tend to confound visual complaints from visual performance assessment with those of the task; thereby explaining the negative first order differences, but also rendering items 1–3 and 10, useless. Accordingly, a subsequent analysis was performed in which the musculoskeletal items only (4–9) were combined into a single scale.

Table 4.6 Experiment 1. Discomfort questionnaire. Mean first order (before–after) and second order (suboptimal–optimal) differences by session half (first/second).

	First half			Second half		
Question	Suboptimal	Optimal	Difference	Suboptimal	Optimal	Difference
Blurred vision	0·62	−2·62	3·24	−1·00	0·08	−1·08
Eye irritation	3·46	−0·15	3·61	2·77	2·23	0·54
Eye fatigue	6·00	2·69	3·31	2·08	6·00	−3·92
Pain in arms	8·15	6·00	2·15	7·54	4·54	3·00
Pain in legs	3·15	3·69	−0·54	5·46	3·15	2·31
Pain in neck/upper back/shoulder	14·85	8·54	6·31 *	12·31	11·31	1·00
Pain in wrist/hands and fingers	6·00	6·15	−0·15	5·92	6·38	−0·46
Pain in lower back	17·15	10·00	7·15 *	9·85	9·39	0·46
Loss of sensitivity in fingers	0·54	0·62	−0·08	0·00	0·00	0·00
Headache	4·39	1·54	2·85	2·23	1·15	1·08

* = significance at $p<0{\cdot}01$

For each subject, average first order differences were summed across items 4–9 within session half (first/second) and condition (optimal/suboptimal). These difference values were then subjected to a 2 × 2 repeated measures analysis of variance.

Results confirm that musculoskeletal complaint levels were greater under suboptimal conditions. The condition factor was significant at the 'liberal' 10% level; $F(1,12)=3{\cdot}09$, $P<0{\cdot}10$. However, neither the session half factor, $F(1,12)=1{\cdot}41$, nor the H × C interaction, $F(1,12)=1{\cdot}83$, approached statistical significance. During the first half of the session, mean first order differences were 8·31 for suboptimal, and 5·83 for optimal conditions. Comparable values for the second half were 6·85 and 5·79. Second order differences (suboptimal – optimal) were 2·48 for the first half, and 1·06 for second half.

Experiment 2

Table 4.7 contains the mean first and second order difference by condition and session half. (Because of a coding error in the data transmission process, only 14 subjects are included in this analysis.) Correlated one-tailed *t*-tests were run on all second order differences. Significant differences in the predicted direction were found for item 3 (eye fatigue), $t(13) = 2{\cdot}14$, $P<0{\cdot}05$; item 8 (pain in lower back), $t(13) = 2{\cdot}43$, $P<0{\cdot}03$; and for item 10 (headache), $t(13) = 2{\cdot}13$, $P<0{\cdot}05$ during the first half of the session. No significant differences were discovered during the second half.

Table 4.7. Experiment 2. Discomfort questionnaire. Mean first order (before–after) and second order (suboptimal–optimal) differences by session half (first/second).

	First half			Second half		
Question	Suboptimal	Optimal	Difference	Suboptimal	Optimal	Difference
Blurred vision	2·86	1·36	1·50	3·71	4·93	−1·22
Eye irritation	1·57	−0·43	2·00	6·14	3·07	3·07
Eye fatigue	6·71	1·00	5·71 *	7·86	4·57	3·29
Pain in arms	8·43	6·14	2·29	13·00	8·07	4·93
Pain in legs	4·14	3·07	1·07	4·86	3·86	1·00
Pain in neck/upper back/shoulder	24·07	17·36	6·71	18·00	15·21	2·79
Pain in wrist/hands and fingers	13·71	9·50	4·21	13·07	13·21	−0·14
Pain in lower back	18·79	6·36	12·43 *	16·86	11·14	5·72
Loss of sensitivity in fingers	4·93	3·57	1·36	5·29	5·36	−0·07
Headache	5·57	0·21	5·36 *	1·79	3·29	−1·50

* = significance at p<0·01

As just discussed, concern about contamination of the visual complaint items by the visual performance procedures led to the discarding of those items, and the combination of only the musculoskeletal items into a single scale. However, since Experiment 2 did not involve visual function testing, it seemed reasonable to utilize the information from the complete 10-item questionnaire in conducting a combined analysis comparable to that of Experiment 1. Accordingly, all items were combined into a single discomfort complaint scale by summation of the average first order differences. These difference values were then subjected to a 2 × 2 repeated measures analysis of variance similar to those described above. Factors were session half (first/second) and condition (optimal/suboptimal).

Results were not as clear-cut as in Experiment 1. The main effect of condition did not achieve even the liberal 10% level of statistical significance; $F(1,13) = 2{\cdot}77$, $P = 0{\cdot}12$. However, although the session half effect did not approach significance, $F(1,13) = 1{\cdot}08$, the C × H effect was significant; $F(1,13) = 3{\cdot}06$; $P<0{\cdot}10$. During the first half, the average combined first order differences were 9·08 for suboptimal, 4·81 for optimal. Comparable values for second half were 9·06 and 7·27. Second order differences were 4·27 for the first half, and 1·69 in the second half. Thus, the significant interaction indicates that the complaint levels were relatively greater in the suboptimal conditions only during the first 2 days of the study.

Final interview

At the conclusion of the 5-day sequence, each subject was given a final interview during which the following questions were posed: 'What did you think was going on? What did you think about the two rooms you worked in?' A simple item analysis of each subject's (unprompted) comments regarding specific features of either optimal or suboptimal workstations was carried out.

In both Experiments 1 and 2, subjects unanimously preferred the optimal workstation, some of them emphatically. Each subject had at least one positive comment regarding the comparative advantage of the optimal over the suboptimal workstation; the majority of which related to the ergonomic chair, worksurface height and copy holder.

In Experiment 1, the only negative comments regarding the optimal workstations came from three subjects, each of whom complained that lighting levels were too low compared with the suboptimal station. In fact, because of glare reduction measures, illuminance in the suboptimal workstation was almost twice that of the optimal stations. These subjects seem to have preferred more light intensity despite its accompanying glare.

In Experiment 2, three subjects described the lighting conditions in the optimal workstation as superior to those in the suboptimal. Recall that we had attempted to make these two stations photometrically equivalent. There seems, therefore, to have been a qualitative lighting difference between the two workstations despite our efforts.

Discussion

In the experimental data-entry simulation in which postural components (working surface height, copy holder, wrist rest, adjustable ergonomic chair and screen height) as well as visual components (lighting fixtures, antiglare screen, contrast levels) are compared (Experiment 1), we find that for the same individuals working under optimal ergonomic conditions results in higher levels of performance and lower levels of discomfort complaints. Thus, we can give a clear affirmative answer to the question posed in the introduction: Did the ergonomic differences between workstations have any effect on the operators working in them?

These positive results also give further credence to the concept, elaborated earlier, of operator efficiency. Presumably, what has occurred in this experiment is that the ergonomic components comprising the optimal workstation allowed operators to maintain more favourable postural configurations over the course of the working sessions, in the biomechanical sense that muscles and limbs were closer to their optimum positions for sustained static loads. Combined with the beneficial effects of glare reduction, these improvements enabled subjects to achieve higher levels of keystroke output and lower levels of complaints.

The conclusion can be supported by the independent replication of this study in Experiment 2. Here the postural conditions from Experiment 1 were repeated, but the lighting conditions were made similar. Since the global ergonomic difference between workplaces in Experiment 2 was not as great as in Experiment 1, we predicted a similar result to Experiment 1 but of smaller magnitude. These predictions appear to have been supported. Using average combined incentive pay as a comparative overall measure of performance, the optimal–suboptimal improvement of 17·5%, while statistically significant, is not as large as the comparable value of 23·3% obtained in Experiment 1. Thus, the lighting/glare control seems to add an additional value of 5·8% in terms of performance improvement.

We can make a similar, if less direct argument from the discomfort questionnaire results. Recall that, in Experiment 1, analysis of the combined scale yielded a significant main effect of ergonomic condition only, whereas in Experiment 2, there was a significant half by condition interaction. We might interpret this as indicating that, when the workstation differences were greater (Experiment 1), the relatively greater level of complaints associated with the suboptimal conditions persisted across both halves of the experiment. On the other hand, in Experiment 2, where the ergonomic contrast was less, the suboptimal–optimal difference in complaint level disappeared during the second 2 days.

In each of the above experiments, the subjective reports of operators at the conclusion of the week's work were clearly favourable towards the optimal workstation. Subjects felt they could work more efficiently and more comfortably under the optimal conditions. It is, perhaps, worth reporting that, at the conclusion of her first 3-h session under optimal conditions, one subject wanted to keep working; she told us that this was the most comfortable workstation she had ever used!

Are there alternative explanations of the above results? Technically, the use of counterbalanced designs, as employed in these studies, is open to the risk of asymmetric transfer effects (Poulton and Edwards, 1979). While an independent groups design would have been preferable, the cost and time involved for the number of subjects required made this solution impractical. However, a subsequent analysis of the performance data in each experiment separately by sequence order (OSSO vs. SOOS) confirmed that the observed differences were likely due to the ergonomic conditions (Dainoff, 1983a,b).

A related critique might call upon the so-called 'Hawthorne Effect'. That is, as indicated by the results of the final interviews, subjects were quite clearly aware of the differences between the two workstations. Might this awareness alone have generated differences in either level of performance and/or willingness to report symptoms? Recall that subjects were under incentive pay conditions, and most were earning substantial bonuses. At one level, it seems simply implausible to argue that not only were the relative biomechanical and visual advantages of the optimal workstation irrelevant in affecting the dependent measures, but that, in addition, when subjects worked in suboptimal conditions, they were deliberately lowering their rate of pay because of their perception of poorer working conditions.

This by no means denies the crucial role of higher order ('psychological' or 'management') factors in ergonomic considerations. On the contrary, the laboratory manipulations utilized in the present studies were attempts to hold these factors constant while isolating the impact of ergonomic variables alone. However, when (perhaps on the basis of the positive results of these and other similar studies), management decides to emphasize ergonomic consideration in workplace design, such higher order factors will become essential in successfully implementing an ergonomic programme.

In particular, the operational complexity of a fully-adjustable ergonomic workstation must not be minimized. The operator may be confronted with a chair with as many as nine separate adjustments and a workstation/copy holder/monitor with a similar number. In the controlled environment of the laboratory, these adjustments are fine-tuned for the subject by the experimenter. However, in the workplace, the effective use of such adjustments depends on the knowledge and motivation of the user (who is no longer a 'subject'). Not only must the operation of the controls themselves be understood, but the underlying rationale — to provide postural efficiency/comfort — must be explained at some level or the adjustability will simply not be utilized.

This is particularly important in view of more recent conceptions of seated work posture. At the time of the initiation of this research (1980), optimal posture was defined with respect to the so-called 90° upright position (trunk and legs at 90° to the horizontal; thighs and feet horizontal). However, work by Grandjean *et al.* (1983), and Mandal (1981) presented alternative views of optimal posture in terms of, respectively, backward leaning and forward leaning positions; and suggested chairs/workstations be designed to support these postures. Dainoff and Mark (1987) found both postures effective depending on task constraints, and argued that a fully ergonomic workplace should afford both postures. This approach, of course, requires a well-trained and informed user, thereby placing a major implementation burden on management. However, such an effort ought to be ultimately well worth while.

Acknowledgements

The author is indebted to Dr Michael Smith, Chief of the Motivation and Stress Research Section, Applied Psychology and Ergonomics Branch, Division of Biomedical and Behavioral Sciences, National Institute for Occupational Safety and Health at the time this research was conducted, for his guidance and support, and to L. Fraser, B.J. Taylor and D. Secrest who devoted extraordinary time and effort in accomplishing this research.

References

Coe, J. B., Cuttle, K., McClellon, W. C., Warden, N.J. and Turner, P.J., 1980, *Visual Display Units*. Report W/1/80 (Wellington: New Zealand Department of Health).

Dainoff, M., 1982, Occupational stress factors in visual display terminal (VDT) operation: a review of empirical research. *Behavior and Information Technology*, **1**, 141–176.

Dainoff, M. J., 1983a, *Ergonomic Comparison of Video Display Terminal Workstations I: Effects on Operator Performance, Physical Discomfort, and Visual Function* (Cincinnati, OH: National Institute for Occupational Safety and Health).

Dainoff, M. J., 1983b, *Ergonomic Comparison of Video Display Terminal Workstations III: Effects on Postural Manipulation* (Cincinnati, OH: National Institute for Occupational Safety and Health).

Dainoff, M. J., 1984, Ergonomics of office automation — a conceptual overview. In: *Proceedings of the 1983 International Conference on Occupational Ergonomics*, Vol. 2: Reviews, edited by M. Matthews and R. D. G. Webb, pp. 72–80 (Toronto: Human Factors Association of Canada).

Dainoff, M. J. and Dainoff, M. H., 1986, *People and Productivity: A Manager's Guide to Ergonomics in the Electronic Office* (Toronto: Carswell). (Originally published by Holt, Rinehart and Winston.)

Dainoff, M. J. and Mark, L. S., 1987, Task and the adjustment of ergonomic furniture. In: *Work with Display Units*, edited by B. Knave and P. -G. Wideback (North Holland: Elsevier).

Gould, J. D. and Grischkowsky, N., 1983, Doing the same work with hardcopy and with CRT computer terminals. *Human Factors*, **26**, 323–338.

Grandjean, E., 1979, *Fitting the Task to the Man, An Ergonomic Approach* (London: Taylor and Francis).

Grandjean, E., 1980, Ergonomics of VDUs: review of present knowledge. In: *Ergonomic Aspects of Visual Display Terminals*, edited by E. Grandjean and E. Vigliani (London: Taylor and Francis).

Grandjean, E., Hunting, W. and Piderman, M., 1983, VDT workstation design: preferred settings and their effects. *Human Factors*, **25**, 161–175.

Kroemer, K. H. E. and Price, D. L., 1982, Ergonomics in the office: comfortable workstations allow maximum productivity. *Industrial Engineering*, **14**(7), 24–32.

Mandal, A. C., 1981, The seated man (*Homo sedans*), the seated work position. Theory and practice. *Applied Ergonomics*, **2**(1), 19–26.

McCormick, E. J. and Ilgen, D., 1980, *Industrial Psychology*, 7th edn (Englewood Cliffs, NJ: Prentice-Hall).

National Research Council, 1983, *Video Displays, Work, and Vision* (Washington, DC: National Academy Press).

Poulton, E. C. and Edwards, R. S., 1979, Asymmetric transfer in within-subjects experiments on stress interactions. *Ergonomics*, **22**, 945–961.

Smith, M. J., Cohen, B. F. G., Stammerjohn, L. and Happ, A., 1981, An investigation of health complaints and job stress in video display terminal operations. *Human Factors*, **23**, 387–400.

Starr, S. J., Thompson, C. R. and Shute, S. J., 1982, Effects of video display terminals on telephone operators. *Human Factors*, **24**, 699–711.

Chapter 5

Postural Load and the Incidence of Musculoskeletal Illness[1]

ARNE AARÅS, ROLF WESTGAARD, EINAR STRANDEN AND STIG LARSEN

Introduction

A high rate of sick-leave, labour turnover and rehabilitation problems during 1967–74 were recorded at Standard Telefon og Kabelfabrik (STK)'s telephone plant in Kongsvinger, Norway. This convinced the management and unions of the need to establish a company intervention programme with stated principles for organization and execution of work environmental projects (Aarås and Westgaard, 1980). The programme provided the opportunity to carry out a study of the occurrence of musculoskeletal illness among workers in situations demanding static muscle work.

In early 1975 several major changes were implemented to improve workplace design, giving the workers greater flexibility in adopting a suitable work posture.

The period of study was from 1967 to 1984 and included 420 subjects representing approximately 700 man-labour years before and approximately 800 man-labour years after the ergonomic redesign.

Although this study was carried out because of problems arising with assembly workers, similar problems are reported and have become the focus of attention concerning workers in offices, especially visual display terminal (VDT) operators. In the conclusion from this paper we suggest that the methodology and recommendations reported here could be utilized in similar projects concerned with studies of the computerized office.

[1] The editors feel that, perhaps, a special comment is in order as to why this report by Aarås and his colleagues, relating to light assembly work in a factory situation, is included in a volume dealing mainly with computerized office work. The biomechanical problems addressed in this report (static loads and constrained postures in sedentary work, upper extremity repetition) are, of course, central to the problems encountered in the modern office workplace. More importantly, perhaps, we feel strongly that the general approach and methods of this study, including the outcome evaluation, represent an excellent model which could be largely adopted for investigation of ergonomic interventions to reduce musculoskeletal disorders in the office workplace.

Method and materials

The general aim of the present investigation was to study the relationship between postural load for a group of workers and the development of musculoskeletal illness related to the time of employment for the same group. Further questions emerged:

1. (a) Does the introduction of workplaces designed on ergonomic principles reduce the postural load?
   (b) To what extent does the reduction of postural load influence the incidence of musculoskeletal illness?
2. Is there a relationship between the postural load and the development of musculoskeletal illness when comparing different work tasks?
3. Is it possible to suggest limits for a safe level of work load?

Our attempt to answer these questions was based on the quantification of both postural load and health effects in order to evaluate a cause–effect relationship between these parameters for groups of female workers in well-defined work situations (Westgaard and Aarås, 1984; Aarås and Westgaard, 1987; Aarås *et al.*, 1988).

The factory and the work tasks

The work situations for the different tasks studied are illustrated in Figures 5.1–5.7. The factory was established in 1967 to produce parts for telephone exchanges. The production was first based on the telephone exchange '8B', followed by 'Minimat'. 'Cableforms' were made and used as back-wiring for the two systems. Two new systems, '10C' and '11B', were introduced in the production about the time of redesigning the old workplaces in 1975.

8B system

The work system referred to as 8B is illustrated in Figure 5.1. The main work task consisted of joining wires to terminals of a vertically mounted frame, 100 cm wide and 40–80 cm high, by using a heavy wrapping gun. (The wire was placed in the gun which was then positioned on to the terminal. Thereafter the wrapping gun spun the wire around the terminal.) The wrapping work was done along one horizontal row and the working time varied between 30 and 90 minutes. The work was then immediately continued on the row above or below the finished row. Thus, the work area was limited to about 1 cm^2 at a time. It was necessary to see small needle-shaped terminals about 5 cm inside the frame, which meant that the angle between the line of sight and the horizontal plane must be small. The metal frame was placed on a table of fixed height. Work on the lower rows necessitated a forward bending posture in order to satisfy the vision requirement. Most workers preferred to keep a seated posture as long as possible. The chair was therefore adjusted upwards when the worker moved on to higher rows until the thighs were pushed against the lower edge of the table surface. The chair had a wooden seat and a wooden back support which

Figure 5.1. Old unadjustable 8B telephone system workplace.

could be regulated forwards and backwards, but gave no support for the lumbar area of the back when bending forward. The chair had no arm rests. It was therefore not possible to support elevated arms while working on the middle rows, although the table acted as support when working on the lower two or three rows. The weight of the electric wrapping gun (about 1000 g) was not counterbalanced and this increased the static load on the muscles in the right shoulder and arm. Work in the upper part of the frame had to be done while standing with the shoulders and arms elevated, resulting in static load on the shoulder and arm muscles.

Early in 1975 the workplace was redesigned. The frame was placed on a hydraulic stand (Figure 5.2) with a vertical movement of 60 cm, which was sufficient to allow both seated and standing postures. All seated work could be done at virtually the same height by adjusting the height of the frame. A lighter, pneumatic wrapping gun (350 g) with a counterbalancing system was also introduced.

In order to reduce static muscle load when working with elevated arms (necessary due to vision requirements), arm rests were mounted on an upholstered, padded chair (Figure 5.2). The height and slope of the arm rests were adjustable and it was possible to turn them towards or away from the body. The arm rests were intended to reduce the static muscle load since the vertical working height at any time was no higher than the height of a horizontal row (5–10 cm). The back rest supported the lumbar area of the back. With the redesigned workplace, most of the work could be done seated using arm rests, since the frame could be lowered all the way until the lower edge touched the thighs. When working on the upper rows, a standing posture was often still necessary. The frame would normally be positioned lower than with the old workplace, but the need for good vision made it still necessary to bend forward,

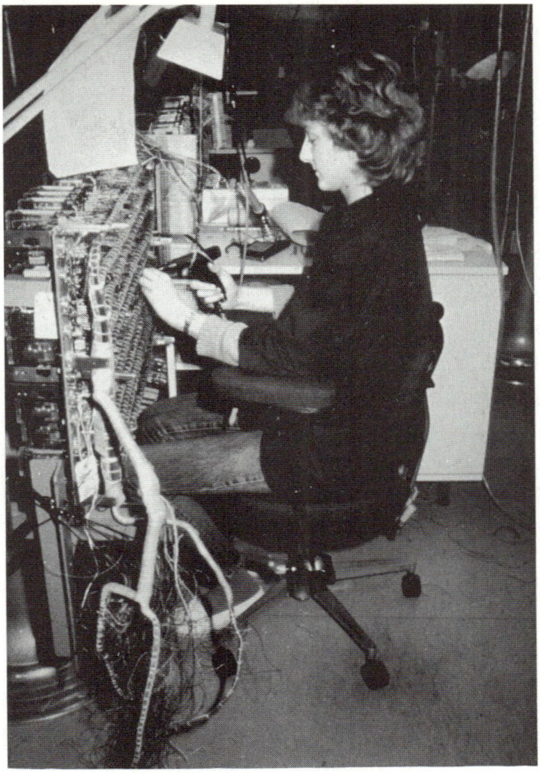

Figure 5.2. New adjustable 8B workplace.

elevate the arms and use an elbow angle of less than 90°. Static muscle load was assumed, nevertheless, to be reduced also when standing due to a somewhat better working height and a lighter wrapping gun.

Cable making

The work situation is illustrated in Figure 5.3. The main work consisted of laying thin insulated wire between nails on a wooden board and sewing the wires together with plastic bands. There were two sizes of wooden boards, 2 × 1.5 m and 4 × 1.5 m. Before 1975, the boards were mounted with fixed centre height, and the slope of the board could be altered in a stepwise manner. This mounting system did not allow any adjustment of the height of the work surface to compensate for variations in the height of the workers and the cableforms. When working on large cableforms, the workers were forced to stretch because the working area was large and too high. Almost all cable making was performed in a standing position due to the large working area. A new mounting system for the wooden board, based on a hydraulic

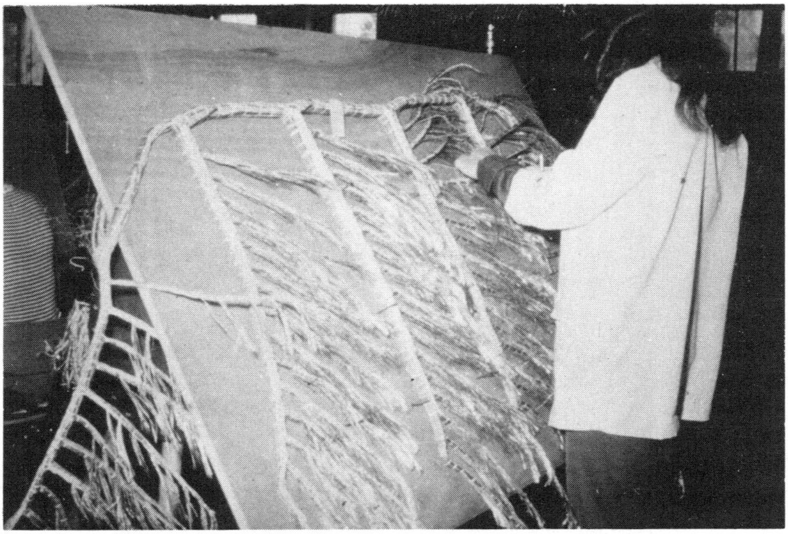

Figure 5.3. Old unadjustable cableform making workplace.

cylinder, was introduced in the last half of 1975 (Figure 5.4). This system allowed continuous adjustment of height and slope of the table. Independent adjustment of height and slope of the table allowed each worker to find his own optimal work posture depending on his height and the size of the cableform, thereby expecting to reduce the strain on neck, shoulders and low back.

The Minimat system

The production work was similar to that described for the 8B system, except the work area was larger. However, the worker could work on one horizontal row after the other, keeping the work height at 5–10 cm. The workplace was adjustable by 60 cm in height. Therefore the worker could adopt a well-balanced sitting posture when working on the lower part of the wooden board.

When working on the middle rows, the board had to be lowered to such an extent that there was no room for the legs below it. This resulted in forward bending with a somewhat twisted posture in order to reach the work area. For approximately 15% of the total work time it was necessary to stand in order to reach the upper rows. The requirement for good vision was similar to that of the 8B system. The forearms were therefore lifted above elbow height resulting in a static load on the shoulder and arm muscles.

Until 1975 the same chair and wrapping gun as for the 8B system was used. In 1975 the new chair with arm rests and the lighter, counterbalanced wrapping gun previously described for the 8B system, was introduced (Figure 5.5).

Figure 5.4. New adjustable cableform making workplace.

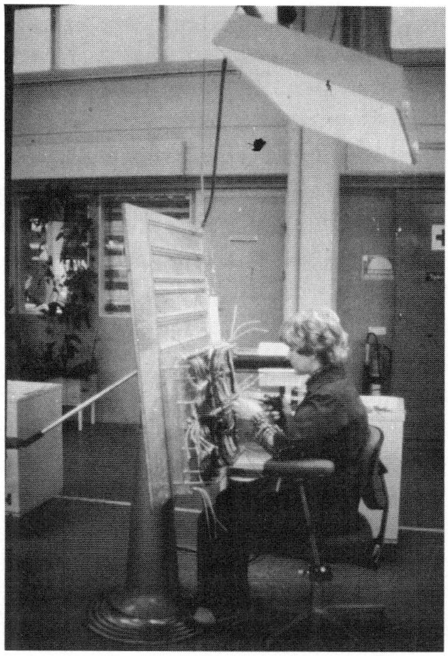

Figure 5.5. New redesigned Minimat system workplace.

Figure 5.6. Adjustable 10C system workplace.

Figure 5.7. Adjustable 11B system workplace.

The 10C system

The work object was a large metal frame, 236 × 60 cm, divided into seven or eight smaller 'sub-frames' which were mounted side by side along the length of the large frame (Figure 5.6). Most of the work consisted of laying long wires between needle-shaped terminals and connecting the wire ends to the terminals with a wrapping gun, without using a pre-produced cableform.

Terminals which were joined together within a sub-frame (about 35% of total work time) were scattered over the whole working area, making the working height the same as the full height of the sub-frame (60 cm). This work was done seated. However, most wires join terminals in different sub-frames so that the wires cross several sub-frames or even the whole length of the large frame. This work (more than 50% of total work time) was done standing. The frame could be lowered or lifted to a suitable height, for both standing and seated work. However, this did not reduce the working height when seated.

The 11B system

The frame was 78 × 100 cm and consisted of two to four 'shelves' mounted on top of each other (Figure 5.7). The work consisted of connecting wire ends to terminals using a wrapping gun. Most connections were done within one shelf, while terminals in different shelves were connected in 20% of the work time. The frame was mounted on a hydraulic stand similar to the 8B frame. However, the high working height and the visual demands (as for the 8B system) meant that the worker had to keep the arms above elbow height, particularly when in the seated position.

The wrapping gun was counterbalanced and chairs with arm rests were used. However, most workers used the arm rests only when the work area was restricted to one of the two lower shelves giving them a seated work position for about 20% of the working time.

Estimation of the risk of health injury

The incidence of musculoskeletal illness was evaluated by using the personal sickness records of each subject. All sick-leave of more than 3 days duration was identified by collecting the medical diagnosis from the local health authorities. The employees were grouped according to their working task (8B, cable making, Minimat, 10C and 11B) from the start of their employment and then year by year until the termination of their employment, transfer to another job, or to the end of the investigation. The musculoskeletal sick-leave was then quantified for each group.

The severity of the health effects was judged from the parameters defined in Table 5.1.

In the quantification of musculoskeletal injury, only load related musculoskeletal illnesses such as tendinitis, myalgia, myotendinitis, lumbago and sciatica were selected. These diagnoses are included because overload on muscles, tendons or supporting tissue is considered to be a major factor in the aetiology of such illnesses.

Table 5.1 Definition of health effects.

Definition	Explanation
1. Development of musculoskeletal sick-leave according to a defined period of employment	The number of workers with one or more recorded periods of musculoskeletal sick-leave as a fraction of total workers at risk within this defined period of employment
2. Median duration of musculoskeletal sick-leave per man-labour years	The median is defined as the observation having as many musculoskeletal sick-leave periods per man-labour years above as below
95% CI of musculoskeletal sick-leave per man-labour years	95% CI of this parameter is defined as an interval which covers the parameter with a probability of 95%
3. Duration of sick-leave	Total duration (days) of each episode of musculoskeletal sick-leave
4. Ratio of musculoskeletal diagnosis	Total number of musculoskeletal diagnoses as a per cent of man-labour years
5. Ratio of workers ill with musculoskeletal sick-leave	Number of workers with musculoskeletal sick-leave as a fraction of total workers employed
6. Musculoskeletal sick-leave (%)	The duration of musculoskeletal sick-leave (days) as per cent of total time of employment (possible working time; Westgaard and Aarås, 1984, 1985) or calendar days (Aarås and Westgaard, 1987; Aarås *et al.*, 1988)
7. Turnover	Number of workers terminating their employment as per cent of average number of people employed

The risk of health injury was also assessed in terms of intensity and duration of pain from the musculoskeletal system, by personal interviews of the workers at the plant in 1984. Pain intensity was quantified on a scale relating pain to work capability (Westgaard *et al.*, 1982). This provided additional information about musculoskeletal complaints which had not led to a period of sick-leave. The duration of painful symptoms was quantified according to the standardized Nordic questionnaires for the analysis of musculoskeletal symptoms (Kuorinka *et al.*, 1987).

Further information on a point prevalance of such illness was obtained by examining the workers clinically for signs of musculoskeletal illness primarily concerning ailments of musculo-tendon organs. This examination was carried out just after the pain interview (Aarås, 1990). The examination consisted of a systematic investigation of the musculoskeletal system such as measurements of the range of passive movements of the neck and head, palpation of muscle spasm and sore spots (trigger points) of m.trapezius (Sheon *et al.*, 1982). Palpation of tendon attachments to the supraspinatus and deltoid muscles was performed with relaxed muscles and against active resistance during the muscle contraction. Tenderness or pain when palpating the tendon attachment were recorded as a positive indication of a

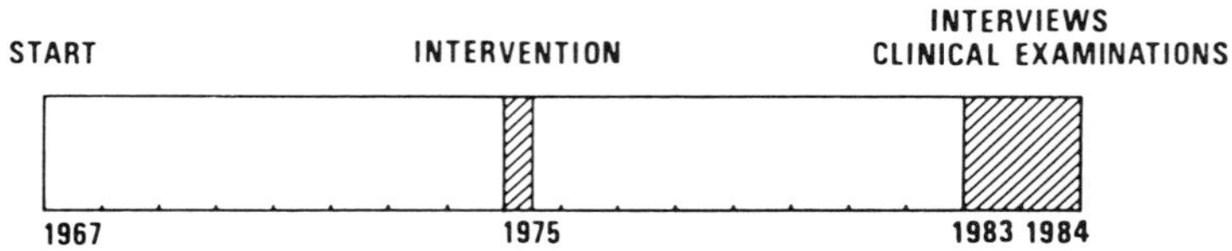

Figure 5.8. Timing of the various activities in the project.

musculoskeletal ailment. Isometric and endurance tests of the shoulder and neck muscles were also performed (Aarås, 1990).

An examination of the spine was carried out according to a screening procedure used by the Department of Orthopaedics, Sahlgren Hospital, Gothenburg, Sweden. The timing of the various activities in this project is illustrated in Figure 5.8.

Subject population

When studying the postural load–musculoskeletal injury relationship, confounding factors such as gender and working hours were controlled by selecting only female workers employed full-time (8 h/day). Mean age at recruitment and mean time of employment are given in Table 5.2 for each work system.

Table 5.2. Subject population parameters.

Work system	Period of time	No. of workers	Mean age at recruitment (years)	Mean duration of employment (years)
8B	1967–74	221	22·1	2·2
Cableform making	1967–74	25	30·9	3·1
Minimat	1970–74	24	24·2	2·1
10C	1975–83	29	24·9	4·3
11B	1974–83	32	24·4	3·6

Postural load

Postural shoulder load was quantified by recording electromyography (EMG), using surface electrodes (Westgaard and Aarås, 1985; Aarås and Westgaard, 1987). The load on m.trapezius (descending part) was selected as an indicator of the load on shoulder and neck area since this muscle provides the main lift for the shoulder girdle, and is important for stabilization of the scapula during arm movements. Also, many subjects reported symptoms located in the body and tendons of the trapezius which were confirmed during medical examination (Aarås, 1990). Postural load was also assessed by recording the movements of the upper arm in the shoulder joint

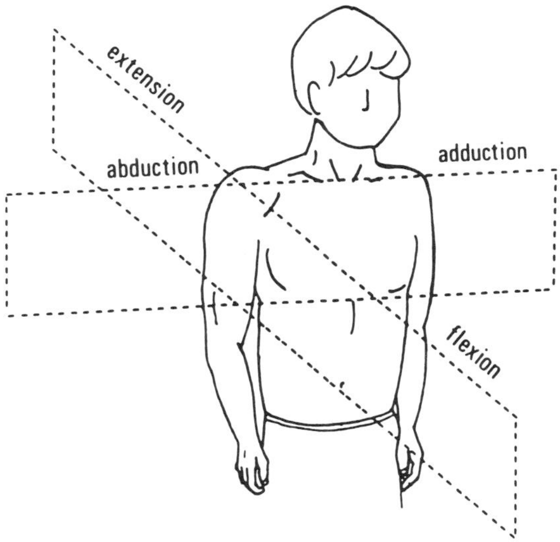

Figure 5.9. The flexion/extension and abduction/adduction of the upper arm in the shoulder joint.

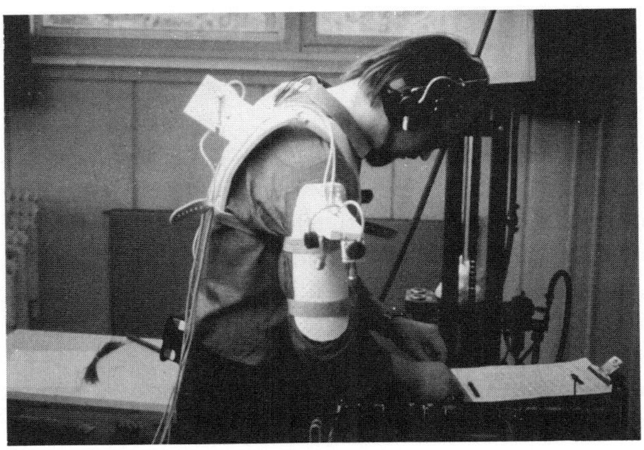

Figure 5.10. The 'physiometer' with angle transducers attached to the head, upper arm and back.

(glenohumeral joint) in terms of flexion/extension and abduction/adduction (Figure 5.9) as well as flexion/extension of the head and back using pendulum potentiometers (Figure 5.10). These postural angles were measured in terms of deviations from a reference body position. This was defined as a standing position with balanced, neutral upright head and trunk posture, relaxed shoulders and both arms hanging relaxed down the body.

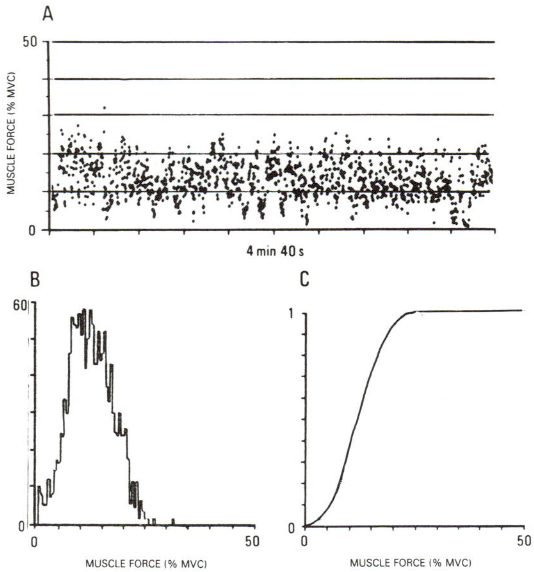

Figure 5.11. (*A*) *Muscle load on the upper trapezius. Each point in the recording gives the mean load in 0·2-s intervals, as the percentage of maximal voluntary contraction* (*% MVC*). (*B*) *Points* (*0·2-s duration*) *are ranked according to muscle force* (*% MVC*) *on the* x-*axis. The number of points are given on the* y-*axis.* (*C*) *The cumulative amplitude probability distribution function of the recording.*

The quantitative EMG technique and the recording of postural angles are described, and the limitations of the methods are discussed, in detail elsewhere (Aarås, 1987; Aarås and Stranden, 1988; Aarås *et al.*, 1988; Westgaard, 1988). However, the main steps in the analysis of signals representing the EMG activity and postural angles are detailed below.

Figure 5.11A shows the quantification of the EMG signal from a 280 s recording from the descending part of the m.trapezius. The analysis is based on digital, full wave rectification and integration of the EMG signals over 0·2-s intervals, resulting in discrete values which are a measure of average electrical muscle activity over this interval. These values were used to estimate the force developed by the trapezius muscle, as a percentage of maximal voluntary contraction (% MVC) of the same muscle, following a calibration procedure with simultaneous recording of force and EMG activity at varying levels of muscle force (Aarås and Westgaard, 1987). Thus, the points in Figure 5.11A represent the muscle force over 0·2-s intervals.

A quantitative analysis of muscle load during this recording was obtained by ranking the interval estimates (0·2-s duration) according to muscle force (Figure 5.11B) to produce a cumulative amplitude probability distribution function (Figure 5.11C) (Jonsson, 1982; Aarås and Westgaard, 1987). The muscle force given by this curve indicates the time fraction of the recording period with the force lower than or equal to a given level. Static force level is defined as the level of muscular contraction corresponding to probability level 0·1, i.e. muscle load is higher than this level for 90% of the recording period. Peak load is similarly defined as the load corresponding to probability level 0·9, i.e. the muscle has a lower force level in 90% of the recording period. Probability level 0·5 defines the median level of contraction.

The analysis of postural angles was carried out according to the same procedure as for the EMG signals.

Analysis of low trapezius activity and low shoulder moment

For the development of musculoskeletal illness, both the muscle load, quantified in terms of the cumulative amplitude distribution function, as well as the number and distribution of time periods with low levels of muscular activity, may be important factors. Therefore, an analysis of such low levels of activity was performed by choosing a level of muscle force and counting the number of periods below this level from the digital, full wave rectified and integrated EMG signal (Figure 5.12A). Correspondingly, for analysis of low levels of postural angles, we calculated the number of activity levels within a window close to the defined reference body position (Figure 5.12B). The dynamic pattern of the work was assessed by counting the number of crossings of the upper arm of the position which this arm kept for 50% of the recording time, i.e. the probability of '0·5' of the cumulative distribution curve of flexion/extension.

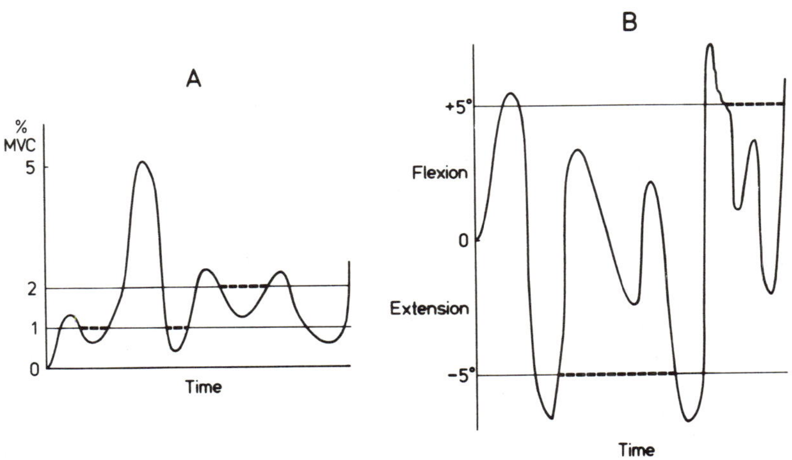

Figure 5.12. (A) Examples of time periods of muscular activity below 1 and 2% MVC are indicated by broken lines. (B) Two time periods of postural angles of the upper arm in the sagittal plane (between +5° (flexion) and −5° (extension)) are shown by broken lines.

Statistical methods

All new results presented in this chapter (Tables 5.3–5.5) are expressed as medians and total ranges. Due to the limited number of subjects included in each group, no significance tests were carried out. Concerning results referred to from previous papers, the statistical methods are discussed in Aarås and Westgaard (1987).

Results

The influence of the ergonomic redesign on the trapezius load and the postural angles in the glenohumeral joint and the development of musculoskeletal illness

The results presented in this section are a summary of publications by Aarås and Westgaard (1987), Aarås *et al.* (1988) and Westgaard and Aarås (1985).

Trapezius load and postural angles in the glenohumeral joint

For workers at the 8B system, the effect of the ergonomic redesign was a considerable statistically significant reduction in static trapezius load as a median group value from the old to the new work situation, both with sitting (4·3 vs. 1·9% MVC) and standing (5·2 vs. 1·2% MVC) postures.

The mechanical shoulder load assessed by measuring flexion/extension in the shoulder joint showed no significant difference in flexion angles between workers in the original and redesigned work situations at the 8B system. However, workers with high values of flexion in the original work situation recorded a considerable reduction of these angles at the adjustable workplaces.

At the cableform making, the trapezius load showed no significant difference when comparing the same work task in the old and the new work situation. However, at the time of the ergonomic redesign, the main work at cableform making was switched from producing small forms (about 1·5 × 1·0 m) which required relatively high static load (median group value, 5·5% MVC; range, 0·7–6.2% MVC), to larger forms (about 3 × 1·5 m, Figures 5.3 and 5.4) which was a less demanding task (median group value, 0·8% MVC; 0·5–4.2% MVC), independently of the redesign of the workplaces.

At the Minimat system, the number of EMG recordings was limited due to the cessation of manufacture of the system at the time the measurements were performed. The inter-individual variation of static muscle load was considerable and precludes drawing general conclusions.

The new workplaces for the 10C and 11B systems were designed on the basis of the same ergonomic principles as used when redesigning the workplaces of the original systems. Workers at the 10C and 11B showed a much lower median group static muscle load for the predominant standing work position, 1·0 and 1·3% MVC, respectively, compared with those workers employed at the 8B system in the old work situation.

More than 75% and 50% of the work at the 11B and 10C systems, respectively, was performed in a standing position which created less load on the shoulder compared with a seated posture. The static trapezius load as a median group value compared for sitting and standing position on the 10C and 11B, was 3·8 vs. 1% MVC and 4·1 vs. 1·3% MVC, respectively. These values, particularly for the standing position, were much lower than the corresponding values for workers at the 8B system in the original work situation.

Musculoskeletal sick-leave

Musculoskeletal sick-leave was reduced after the ergonomic redesign when comparing workers at the 8B system who had been employed for at least 1 year both in the original and the redesigned work situations in order to have a sufficient period of observation to assess the effect of the workload on the health of the workers. The same subjects were compared to eliminate individual differences in general health condition and in ability to tolerate prolonged muscle load. To assess changes in the health condition, only subjects with at least one recorded musculoskeletal sick-leave were included in the comparison. The sick-leave was calculated as the mean number of days absent from work per year due to musculoskeletal illness for each worker. The median and 95% confidence interval (CI) were determined from the resulting distribution (Aarås and Westgaard, 1987). According to this calculation, periods of sick-leave were reduced (as a median duration per man-labour years with 95% CI) from 22·9 days (4·4–50·8) in the old situation to 1·8 days (0–34·4) in the new, a difference which is statistically significant ($p<0\cdot02$).

This reduction was obtained in spite of a longer time of employment (mean, 4 years) per worker and a higher mean age (4·4 years) of the workers in the period after the redesign of the workplaces. The total work time for the group was 66 years before and 59 years after redesign, a difference which is not statistically significant.

Cableform making showed also a statistically significant reduction in musculoskeletal sick-leave as a percentage of possible workdays, from 4·7 to 1·6%, and the number of such absences as a fraction of man-labour years was also reduced from 30·6 to 14·3% when comparing the periods before and after the ergonomic redesign.

The effect of the ergonomic redesign, as such, was also assessed by comparing all workers employed at the old systems before the redesign with workers recruited after 1974 and employed only at the redesigned systems. The latter group of workers had a much higher probability of *not* getting a musculoskeletal sick-leave at any time during the first 2 years of employment than workers at the old systems. After 2 years of employment, the difference was higly significant ($p\ll0\cdot01$).

However, a positive effect of the ergonomic redesign, as such, was also provided by a sharp interruption of the increasing incidence of musculoskeletal sick-leave which coincided with the ergonomic redesign.

The reduction in musculoskeletal sick-leave was not due to reduced workload in terms of production demand. By using an estimate of standard time for a well-defined unit of work as a basis, the productivity was on average higher in the period 1975–82 compared with 1969–74.

Other effects of the ergonomic redesign

Another indication of improved environmental working conditions was a statistically higher significant reduction of labour turnover from 30·1% (1967–74) to 7·6% of total man-labour years (i.e. average number of workers) in the period 1975–82 ($p<0{\cdot}001$).

The reduction in musculoskeletal sick-leave and labour turnover, when comparing the periods prior to and after the ergonomic redesign, are also discussed from a cost–benefit point of view by Spilling *et al.* (1986). A total investment in the work environment programme of about US$57 000, produced a significant reduction in direct costs to STK of about US$490 000 in the period 1976–88. The investment and savings were both calculated in present values (1976 value).

Relationship between trapezius activity, postural angles and pattern of movement of the upper arm in the glenohumeral joint and the development of musculoskeletal sick-leave when comparing different work tasks

This section presents new information on the importance of low trapezius activity and low shoulder moment regarding the incidence of musculoskeletal sick-leave. The amount of dynamic work also seems to influence such illness. Further details of the trapezius load and postural angles related to musculoskeletal sick-leave have already been discussed by Aarås and Westgaard (1987) and Aarås *et al.* (1988).

Trapezius load

Postural load clearly influenced the development of musculoskeletal sick-leave when considering the work situation before and after the ergonomic redesign of the workplaces. Further support for a postural load–musculoskeletal injury relationship was obtained by comparing workers at the 8B system, cableform making and the Minimat systems in the old work situation with workers employed at the redesigned 10C and 11B systems.

The median value of static trapezius load was similar for groups of workers employed at the 8B, cableform making and the Minimat systems in the old work situation; 4·3–5·5% MVC regardless of sitting or standing work positions. For the redesigned systems 10C and 11B, the group median value of the static trapezius load was only 1 and 1·3% MVC, respectively, when considering the predominant standing work position.

Postural angles in the glenohumeral joint

The shoulder loading in terms of upper arm flexion/extension for workers at the 8B system showed no significant difference in the values of median flexion compared with workers at the 10C and 11B systems. However, workers at the 10C and 11B systems had less upper arm flexion compared with those who worked with high flexion angle at the original 8B system. The median angle of flexion/extension varied between −14° and 8° for the 10C and −11° and 15° for the 11B systems, respectively. Those workers at the 8B system with high median flexion, recorded angles between 26° and 44°. The arm abduction/adduction was low and rather invariant for all the three systems, indicating that the main arm movement was in the sagittal plane.

However, the static, median and peak angles indicated clearly a different pattern of movement of the upper arm in the sagittal plane for most workers at the 10C and 11B systems relative to the 8B system. For the latter system, the workers showed only flexion in the shoulder regardless of working position, while most workers at the 10C and 11B systems performed wiring with both flexion and extension of the upper arm.

An external load in the hand and requirement for precision of movements will increase the trapezius load independently of the postural angle of the arm. The wrapping gun was about 0·7 kg heavier for workers in the original work situation at the 8B system compared with the 10C and 11B systems, where workers used a wrapping gun of about 0·3 kg.

The requirement for precision of movement was the same for the three systems.

Low trapezius activity

Since we have established a relationship between static trapezius load and the development of musculoskeletal illness (Aarås, 1987), any positive indications of low muscle activity (low levels) could be assumed important in its prevention. It can be argued on a physiological basis that pauses or short breaks in the muscle load pattern is necessary for recovery of muscle function. Therefore, an analysis of EMG activity recorded from m.trapezius was undertaken to establish any evidence of low muscle activity, to support the overall hypothesis.

Table 5.3. Low levels of EMG activity in the upper trapezius muscle.

Work system	Work operation	No. of recordings	Activity levels (% MVC)	No. of levels per minute		Duration of level (s/min)	
				Median	Range	Median	Range
8B	Wiring sitting	7	<1	0·3	0 – 0·5	0·4	0 – 1·2
	Wiring standing	6	<1	0·5	0·1– 2·6	1·7	0·1– 6·3
	Wiring sitting	7	<2	0·4	0 – 2·2	1·6	0 – 2·9
	Wiring standing	6	<2	1·4	0·3– 7·5	3·1	0·3–20·2
10C	Wiring sitting	8	<1	1·2	0 – 2·2	3·0	0 – 5·7
	Wiring standing	6	<1	0·9	0·1– 9·5	0·9	0·1–38·0
	Wiring sitting	8	<2	1·7	0·1– 3·6	5·3	0·1–11·2
	Wiring standing	6	<2	2·9	1·5–10·6	3·9	1·7–40·4
11B	Wiring sitting	2	<1	0·6	0·2– 1·1	1·1	0·2– 2·1
	Wiring standing	4	<1	2·0	0·6– 3·3	3·0	1 – 5·6
	Wiring sitting	2	<2	0·6	0·3– 1·0	1·3	0·3– 2·3
	Wiring standing	4	<2	3·1	1·6– 5·9	7·6	2·6–12·4

We counted the number of times that the EMG activity level was below 1 and 2% MVC. Table 5.3 presents the number of periods below 1 and 2% MVC as well as their total duration in seconds per minute (median value with range). This table shows that there were fewer 'low levels'/minute and that the total duration of 'low levels' was less for the predominantly seated operators at the 8B system than for the mainly standing operators of the 10C or 11B systems, both at 1 and 2% MVC.

A comparison of posture within systems also confirmed that sitting work mostly led to fewer 'low levels' than standing (median value at 1% MVC). At the 2% MVC level, all sitting postures recorded fewer 'low levels' than the corresponding standing posture. A similar comparison of posture within systems showed again that the total duration of low level activity was generally shorter for sitting postures. The number and total duration of 'low levels' also shows less variability when sitting compared with standing. This was especially pronounced for seated workers at the 8B and 11B systems.

These results show that the 8B workers had a higher static loading which produces fewer breaks in the high activity periods compared with workers at the 10C and 11B systems.

Low shoulder moment

We also assumed that low postural angles (flexion/extension in the shoulder joint) between +5° and −5° was associated with low moment about the shoulder joint in spite of a maximum flexion of 90° in the elbow joint and a hand tool of about 0·3–1·0 kg. Table 5.4 shows that for the predominant working position, sitting at the 8B system, the number of low angles in the shoulder joint, as a median value, was fewer than for the mainly standing workers at the 10C or 11B systems (0·3 vs. 9·8 and 3·3).

For each system the seated position led, without exception, to fewer angles as a median value within the defined limits than standing. This was also true for the total duration of low angles per minute (0·5 vs. 13·7 and 5·6) for 8B, 10C and 11B workers, respectively. Likewise the variability was high for number and duration of low angles for all systems.

Table 5.4. Activity levels of flexion/extension in the shoulder joint between +5° and −5°.

Work system	Work operation	No. of recordings	No. of activities per minute		Duration of activities (s/min)	
			Median	Range	Median	Range
8B	Wiring sitting	6	0·3	0 – 7·0	0·5	0 –36·3
	Wiring standing	6	2·6	0 – 5·7	7·2	0 –20·6
10C	Wiring sitting	3	4·6	0·3– 7·0	16·6	2·0–19·7
	Wiring standing	3	9·8	1·4–13·1	13·7	0·3–34·9
11B	Wiring sitting	2	2·2	0·1– 4·3	7·6	0·1–15·1
	Wiring standing	4	3·3	2·4– 5·2	5·6	4·6–12·5

The pattern of movement of the upper arm in the sagittal plane

The data on postural angles in the sagittal plane shows even differences in the amount of static or dynamic work between the systems. The work pattern was assessed by the number of crossings at probability 0·5 (i.e. the level the postural angle had for 50% of the recording time). Table 5.5 shows the number of crossings per minute (median value with range). Considering the predominant sitting work position for the 8B system and the mainly standing position for workers at the 10C and 11B systems, the median value was 4·2 (8B) which is lower than 7·4 (10C). Range of movement for 80% of the recording time as a median value is larger for workers at the 10C compared with operators at the 8B system. Number of crossings for workers at the 11B is almost the same as for the 8B system. However, the range of movements for

Table 5.5. Movement of the upper arm in the shoulder joint in the sagittal plane.

Work system	Work operation	No. of recordings	No. of crossings per minute		Range[a] of movements (degrees)	
			Median	Range	Median	Range
8B	Wiring sitting	6	4·2	1·1– 8·6	20·7	17·5–33·7
	Wiring standing	6	3·4	2·1– 4·9	28·5	17·5–35·2
10C	Wiring sitting	3	6·1	4·6– 6·3	23·1	19 –28·4
	Wiring standing	3	7·4	3·5–11·9	26·9	9·5–28·3
11B	Wiring sitting	2	5·0	3·5– 6·5	28·9	18·9–38·8
	Wiring standing	4	3·7	3·1– 4·1	45·3	9·2–50·8

[a]The range of movements is calculated between $p = 0·1$ and $p = 0·9$ (80% of the recording time).

workers at the 11B was more than double the median value of operators at the 8B system, indicating that 11B operators had a dynamic work pattern similar to workers at the 10C system.

Musculoskeletal sick-leave

Operators on the original workstations at the Minimat, cableform making and the 8B systems (1967–74) had a statistically significant higher incidence of musculoskeletal sick-leave than workers at the redesigned 10C and 11B systems (1975–83), when comparing workers employed for 0–2 and 2–5 years ($p<0·01$). For workers employed more than 5 years, a statistically significant higher incidence of musculoskeletal illness was found for workers at the 8B system than for workers at the 10C and 11B systems ($P = 0·05$). The different development of musculoskeletal sick-leave was not due to differences in age at the time of employment at the different systems; the mean age at recruitment was 24·9 years (10C), 24·4 years (11B) and 22·1 years at the 8B system.

In addition, when considering only the development of musculoskeletal illness in the shoulder and neck regions, the workers at the 8B system had a significantly higher incidence of such illness compared with those employed at the 10C and 11B systems (within the time of employment between 2 and 5 years ($p<0\cdot05$)). The overall time of employment was similar for the three groups of workers. Median values varying between 3·1 and 3·9 years, showed no statistically significant difference.

Psychosocial problems and activities outside work may have influenced the development of musculoskeletal illness for the individual and at the group level. However, a coarse quantification of some important psychosocial factors, spare-time activities and living habits of the workers, showed no statistically significant difference with regard to these factors between the groups in 1984 (Aarås, 1990). Other situational variables such as the foremen and the factory layout were also unchanged during the period of investigation (1967–83).

Data indicating limits for a safe level of the workload

There is some difficulty in suggesting limits for a safe level of workload. The workers at the 10C and 11B systems had a median group value of static trapezius load of 1·0 and 1·3% MVC for most of the workday, quantified by the procedure of Jonsson (1982). The median group value of total duration of trapezius load for the operators below 1 and 2% MVC as a percentage of total recording time, was 1·5% (10C) and 5% (11B) (<1% MVC) and 6·5% (10C) and 12·7% (11B) (<2% MVC) (Table 5.3).

The workload assessed by the median flexion/extension in the shoulder joint was less than 15°, and the abduction/adduction was mostly less than 10°. The median group value of total duration of flexion/extension between +5° and −5°, as a per cent of total recording time, was for the 11B and 10C operators 9·3% and 22·8%, respectively. Real breaks in the load pattern in terms of spontaneous pauses varied individually for workers at the 10C and 11B systems and amounted to 30–60 min for each worker, for an 8-h workday.

The development of musculoskeletal sick-leave in the upper part of the body for workers at the 10C and 11B systems, approximated the incidence of such illness, regardless of body location, for a control group of female workers *without* continuous workload. This implies that the trapezius load, as well as shoulder load in terms of flexion/extension and abduction/adduction in the glenohumeral joint, seems to approximate an acceptable level of the workload on the neck and shoulder regions, as reflected by the incidence of musculoskeletal sick-leave.

A discussion of acceptable muscle load must also consider other factors which influence the load on the musculoskeletal system and the incidence of musculoskeletal illness (Kvarnstrøm, 1983). These include the workload outside the plant, psychosocial problems, spare-time activities and living habits (Øgar, 1977; Kogstad and Hartvig, 1978). In the present study, workload at home and psychological problems were the only two factors in the redesigned work situation considered as potential contributors to the overall load on the musculoskeletal system. Considerable work at home and indications of serious psychological problems were reported by

20% of the workers, but a combination of these factors was experienced by only 5% of the workers. Factors considered to strengthen the musculoskeletal system, such as sport and other recreational and leisure activities, were practised only occasionally by less than 35% of the workers. The health consequences of the workload on workers in the redesigned work situation were still considerable. In the last year before the interview (1983), pain was experienced by every fourth worker to an intensity necessitating pauses during working hours. Every fifth worker reported that such pain did not disappear during pauses and made it difficult to carry out work. The duration of such periods with pain was more than 30 days in total over the last year before the interviews. These results are supported by the fact that every third worker reported continuous tenderness or pain after cessation of the isometric and endurance muscle tests (Aarås, 1990).

Discussion

The hypothesis postulating a relationship between the level of postural load and the incidence of musculoskeletal illness, assumes a reduction in the occurrence of such illness following reduction of postural load. The ergonomic redesign in 1975 significantly reduced the static trapezius load for workers at the 8B system, particularly for standing work.

In contrast, flexion in the glenohumeral joint was only reduced for those workers with high angles by adjusting the redesigned work frame. The differential effects of the ergonomic redesign on trapezius load and flexion angles may be explained by the reduction of the shoulder loading due to support of the forearms and reduction of the hand tool weight to almost zero.

In parallel with the reduced postural load when working at the redesigned work stands, the same subjects' health was improved in terms of reduced musculoskeletal sick-leave for the 8B system. Further support for the hypothesis was found by studying cableform makers. The postural load was significantly reduced for many of the workers by transferring them from small cableforms, requiring relatively high static trapezius load, to less demanding work at the large cableforms.

The importance of reducing the postural load by using workplaces designed according to ergonomic principles is supported by a significant reduction of the probability of developing musculoskeletal related sick-leave. This was only observed for those employed at the redesigned workplaces, compared with those who worked at the original work stands at any time during the first 2 years of employment. Both this reduction in musculoskeletal illness as well as in the labour turnover rate were shown to be very profitable for the company.

A relationship between the level of static trapezius load and the incidence of musculoskeletal illness was also found when comparing workers at different systems, i.e. the 8B system vs. 10C and 11B systems. The work stands of the 10C and 11B systems were designed according to ergonomic principles almost from the start of manufacturing of these systems. Both the static trapezius load and the development of musculoskeletal sick-leave were lower for workers employed at these systems compared with the original situation at the 8B system.

Thus, the data provide evidence for a relationship between the incidence of musculoskeletal sick-leave and static trapezius load between groups of female workers. However, the great variation in static trapezius load within each group of workers and the length of the 95% confidence interval of musculoskeletal sick-leave per man-labour years within each group, indicate clearly that such a relationship is only valid when considering the median value for the group (Aarås and Westgaard, 1987). The large difference recorded in workload within the group doing the same task may be due to differences in speed of movements by the subjects, differences in work technique or stressful stimuli from mental problems (Westgaard and Bjørklund, 1987). The great variability of time from recruitment until the subjects take their first musculoskeletal sick-leave within the group, may be due to individual differences in general health condition and ability to tolerate prolonged muscle load, as well as individual differences in muscle load. As an alternative way of analysing the EMG-recordings of trapezius load pattern, the assumption was made that periods with very low activity in the work pattern may be a load variable influencing the development of musculoskeletal illness. This hypothesis was based on data from Lehmann (1962) who showed that both the length of the pause as well as the length of the period of work activity, may influence the development of fatigue. Furthermore, Rohmert (1973) reported that optimum performance was obtained by giving the operators short rest pauses. Micropauses of 10–15 s every tenth minute over a working day led to data-entry operators perceiving less fatigue in the shoulder/neck than on an ordinary workday without such pauses (Ehnstrøm, 1981). Therefore, we examined the number and duration of periods with trapezius activity below 1 and 2% MVC. We based the choice of 1 and 2% MVC level on the assumption that static trapezius load of 1–2% MVC seemed to approximate a tolerable trapezius load for most of the working day as evidenced by the incidence of musculoskeletal illness (Aarås and Westgaard, 1987). Both the number and the total duration of levels below 1 and 2% MVC as a median value were less for the predominant work position, sitting at the 8B system compared with the mainly standing operators at the 10C and 11B systems. These results indicate that workers at the latter systems had lower trapezius activity as well as longer duration of such 'low levels' of activity compared with operators with the predominant posture at the 8B system. Standing seemed more preferable than a sitting position in terms of load on the trapezius muscle, showing a clear tendency to have more 'low levels'; such levels were of longer duration compared with the sitting position. These results were consistent with the recording of a lower static trapezius load as group median for standing than for sitting work (Aarås and Westgaard, 1987).

When quantifying shoulder load in terms of postural angles, a low level of shoulder load can be defined as a relatively low load moment about the shoulder joint, assessed in terms of an arm position between +5° (flexion) and −5° (extension). The results showed clearly that workers at the 10C and 11B systems in the predominant standing position had considerably more of such low angles, which were also of longer duration, compared with the seated operators at the 8B system.

The work pattern, in terms of the amount of dynamic or static work, may be another load variable influencing the development of musculoskeletal illness. Dynamic work pattern increases work capacity compared with static work

(Hagberg, 1981). Furthermore, the importance of a more dynamic work pattern is documented by Kilbom *et al.* (1986), who found that the number of flexing or abducting movements of the upper arm per hour were negatively related to neck, shoulder–neck angle and shoulder symptoms. A more dynamic work pattern for operators at the 10C and 11B systems may have contributed to less discomfort, compared with workers at the 8B system (Table 5.5).

Although this study was carried out among assembly workers, it is suggested that the results are also valid when load is assessed in other work situations, such as VDT work. The essential prerequisite is that the same methods for assessing the postural load are applied and that their limitations are taken into account (Aarås, 1987). Concerning muscle load, it is possible to compare such load between different types of work, between subjects and on different occasions when a calibration of the myoelectric signal amplitude to indicate muscle force is performed (Westgaard, 1988). Therefore these findings are also of general interest concerning work in, for example, a computerized office.

There is some difficulty in suggesting a threshold level for acceptable mechanical load on the shoulder. It is necessary to consider both intensity and duration of this load, as well as the length of necessary rest pauses between periods with prolonged mechanical load. The basis for a suggestion of an acceptable workload should be quantifiable health parameters such as the incidence of musculoskeletal illness among any given group of workers. Further, it is necessary to take into account other factors influencing the incidence of this illness, such as age and gender, working hours per day and total time of employment.

Shoulder load experienced at the 10C and 11B systems seemed to approximate an acceptable level, as development of musculoskeletal sick-leave for those workers due to shoulder injuries related to time of employment was approximately the same as for a control group of female workers without a continuous workload (Westgaard and Aarås, 1984; Aarås and Westgaard, 1987). The components of the workload for operators at the 10C and 11B systems included static muscle load, low level of activity together with spontaneous pauses, as well as muscle load from activities outside the plant and loads which may be generated from serious psychological problems.

It is suggested that, based on this work, an acceptable limit for static trapezius load is approximately 1% MVC. This is also supported by Westgaard *et al.* (1986), who have shown that a static trapezius load of 1% MVC is too high if the working day exceeds 8 h. Service workers on the North Sea platforms, working shifts of 12 h/day for 14 days, recorded a median static trapezius load of 1% MVC, and an incidence of musculoskeletal sick-leave which was considerably higher than for workers at the 10C and 11B systems.

However, low levels of activity may also influence workload. The trapezius load as a group median for the 10C and 11B workers was below 1% MVC for less than 4% of the working day. Spontaneous pauses (i.e. rest pauses from work), varied individually for each worker between 6 and 12% of the working day. The influence of such spontaneous pauses on reducing the absenteeism and increasing the output of work are discussed by Grandjean (1980).

Other factors considered to influence the prevalence of musculoskeletal complaints such as a considerable workload outside the plant and serious psychological problems were reported by less than a fourth of the workers. Nonetheless, the symptoms and signs of musculoskeletal complaints observed during the clinical examination of the workers at the 10C and 11B systems suggested that the workload is higher than acceptable. Serious pain both in intensity and duration was reported by about every fifth worker during the last year before the interviews (Aarås, 1990).

The results from this study suggest that the level for maintained static muscle work, proposed by Jonsson (1982) of 2–5% MVC is too high. The lower level of acceptable muscle load from these studies may be explained by the sensitivity of our approach to long-term effects associated with the long exposure times used in the present investigation, which is different from the studies by Jonsson of muscular endurance during constrained static and dynamic work.

When considering shoulder moments for the 10C and 11B operators, a median arm flexion of 15°, and median arm abduction of less than 10° indicated the size of these angles for 50% of the recording time. The duration of low postural angles in the sagittal plane ($< +5°$ and $> -5°$) as a median group value was 10% (11B) and 25% (10C) of the work time. These limits represent only a rough indication of an acceptable arm position and the rest period for the arm, but it should be noted that these values are much lower than limits for these angles proposed by Chaffin and Andersson (1984).

General conclusions

This project has included analysis of the postural load and muscular injuries suffered by workers at specific work situations. The postural load was measured in terms of trapezius load as a percentage of maximal voluntary contraction and postural angles in the glenohumeral joint in degrees. Such measurements are load indicators on the shoulder and neck regions (Aarås, 1987; Aarås *et al.*, 1988; Westgaard, 1988). The effect of the factors influencing the postural load as well as the overall results indicating limits for a safe level of workload from the present investigation are also valid when workload is assessed in other work situations, such as the computerized office. The essential prerequisite is that the same methods are applied and their limitations are taken into account.

This study documents that for groups of female workers, a quantitative relationship is established between static trapezius load, in terms of median value across the group, and the development of musculoskeletal sick-leave related to the time of employment. Furthermore, the number and duration of periods with low trapezius activity as group median value were similarly related.

By assessing the postural load on the shoulder by measuring the flexion/extension in the shoulder joint, there was an indication that the health effect of this load was influenced by:

(a) the magnitude of postural angles themselves,
(b) the distribution of the muscle load between subgroups of muscles such as flexors and extensors, giving correspondingly reduced periods of activity for each group.

(c) the number and duration of very low postural angles, and
(d) the dynamic pattern of work.

The relative importance of all these factors on the incidence of musculoskeletal illness cannot be settled as a result of this study. However, the overall effects of these factors were related to the incidence of musculoskeletal sick-leave as a function of duration of employment for the following:

(a) workers at the 10C and 11B systems vs. operators at the 8B system,
(b) workers at the 8B system prior to and after the ergonomic redesign of the workplaces, and
(c) workers cableform making when large and small cableforms were compared.

Finally, as a result of this work some tentative limits are proposed for workload in terms of static trapezius load and flexion/extension in the shoulder joint. It should be noted that these limits underline the importance of reducing the static muscle load to a minimum. Similarly dynamic muscle work, by giving the operators variation in posture and movements, may be beneficial and strengthen the musculoskeletal system.

Consequences for VDT work

The results from the present study, should be considered during the design of VDT work by:

1. allowing a flexibility in working position and movements by individual height adjustment of the workplace both for sitting and standing work. Variety of work movements may distribute the muscle load between subgroups of muscles such as flexors and extensors, giving correspondingly reduced periods of activity for each group. Standing positions reduce the trapezius load compared with sitting. Height adjustability of the keyboard/work area allows the operator to work with relaxed shoulders as much as possible. To fulfil this requirement, the height of the keyboard and the table top should be within the distance between the height of the elbow and thigh (6–7 cm). If this is not possible, an arm rest should be available to support the arm.
2. working with small angles of flexion/abduction of the upper arm in the shoulder joints.
3. combining the various parts of the VDT work such that it allows as much dynamic muscle work as possible, to prevent a fixed working posture. Such a fixed posture increases the risk of static muscle load and thus, reduces the number and duration of low activity levels on the musculoskeletal system.

Acknowledgements

The authors thank the following colleagues for helpful discussion: Ergonomist Donald Anderson, MSC Flemming Bo Jessen, Professor Roland Ørtengren and Dr Jørgen Eklund.

References

Aarås, A., 1987, Postural load and the development of musculo-skeletal illness. *Scandinavian Journal of Rehabilitation Medicine*, Suppl. No. 18.

Aarås, A., 1990, Load-related musculo-skeletal illness — is ergonomic workplace design a sufficient remedy? In: *Work Design in Practice*, edited by C. M. Haslegrave, J. R. Wilson, E. N. Corlett and I. Manenica (London: Taylor and Francis) pp. 30–40.

Aarås, A. and Stranden, E., 1988, Measurement of postural angles during work. *Ergonomics*, **31**(6), 935–944.

Aarås, A. and Westgaard, R. H., 1980, The organization and execution of work environmental projects in a Norwegian industrial company. *Ergonomics*, **23**(8), 707–726.

Aarås, A. and Westgaard, R. H., 1987, Further studies of postural load and musculo-skeletal injuries of workers at an electromechanical assembly plant. *Applied Ergonomics*, **18**(3), 211–219.

Aarås, A., Westgaard, R. H. and Stranden, E., 1988, Postural angles as an indication of postural load and muscular injury in occupational work situations. *Ergonomics*, **31**(6), 915–933.

Chaffin, D. B. and Andersson, G. B. J., 1984, *Occupational Biomechanics* (New York: John Wiley).

Ehnstrøm, G., 1981, *Ett forsøk med mikropauser bland anställda vid postgirots bokføringsavdelningar* (Umeå, Sweden: National Board of Occupational Safety and Health).

Grandjean, E., 1980, *Fitting the Task to the Man* (London: Taylor and Francis).

Hagberg, M., 1981, Electromyographic signs of shoulder muscular fatigue in two elevated arm positions. *American Journal of Physical Medicine*, **60**, 111–121.

Jonsson, B., 1982, Measurement and evaluation of local muscular strain in the shoulder during constrained work. *Journal of Human Ergology*, **11**, 73–88.

Kilbom, Å., Persson, J. and Jonsson, B., 1986, Disorders of cervicobrachial region among female workers in the electronics industry. *International Journal of Industrial Ergonomics*, **1**, 37–47.

Kogstad, O. and Hartvig, P., 1978, Psykiatriske og sosiale faktorer i reumatologiske fysikalsk medisinsk uførhet. (The influence of psychiatric and social factors in disablement benefit of rheumatological illness.) *The Journal of the Norwegian Medical Association*, **98**, 943–945.

Kuorinka, I., Jonsson, B., Kilbom, Å., Winterberg, H., Bæring-Sørensen, F., Andersson, G. and Jørgensen, K., 1987, Standardised Nordic questionnaires for the analysis of musculo-skeletal symptoms. *Applied Ergonomics*, **18**(3), 233–237.

Kvarnstrøm, S., 1983, Occurrence of musculoskeletal disorders in a manufacturing industry, with special attention to occupational shoulder disorders. *Scandinavian Journal of Rehabilitation Medicine*, Suppl. No.8.

Lehmann, S., 1962, *Praktische Arbeidsphysiologie 2: a uppl.* (Stuttgart: Georg Thieme).

Øgar, B., 1977, Pasienter i norsk alempraksis (Patients treated by general practitioners in Norway.) (Universitetsforlaget Oslo, Bergen, Tromsø). In Norwegian.

Rohmert, W., 1973, Problems in determining rest allowances. *Applied Ergonomics,* **4**, 91–95, 158–162.

Sheon, N. P., Moskowitz, R. and Goldberg, V., 1982, *Soft Tissue Rheumatic Pain: Recognition, Management, Prevention* (Philadelphia: Lea and Febiger).

Spilling, S., Aarås, A. and Eitrheim, J., 1986, Cost–benefit analysis of work environment investment at STK's telephone plant at Kongsvinger. In: *The Ergonomics of Working Postures*, edited by N. Corlett, J. Wilson and I. Manenica (London: Taylor and Francis), pp. 380–397.

Westgaard, R. H., 1988, Measurement and evaluation of postural load in occupational work situations. *European Journal of Applied Physiology*, **57**, 291–304.

Westgaard, R. H. and Aarås, A., 1984, Postural muscle strain as a causal factor in the development of musculo-skeletal illnesses. *Applied Ergonomics*, **15**(3), 162–174.

Westgaard, R. H. and Aarås, A., 1985, The effect of improved workplace design on the development of work-related musculo-skeletal illnesses. *Applied Ergonomics*, **16**(2), 91–97.

Westgaard, R. H. and Bjørklund, R., 1987, The generation of muscle tension additional to body posture. *Ergonomics*, **30**, 911–923.

Westgaard, R. H., Jansen, T. H. and Aarås, A., 1982, Development of musculo-skeletal illnesses among female workers in work situations required static muscle load. (National Institute of Occupational Health, Internal Report). In Norwegian.

Westgaard, R. H., Wærsted, M., Jansen, T. and Aarås, A., 1986, Muscle load and illness associated with constrained body postures. In: *The Ergonomics of Working Postures*, edited by N. Corlett, J. Wilson and I. Manenica (London: Taylor and Francis), pp. 5–18.

Chapter 6

Towards Optimizing Physical Activity in VDT/Office Work

JØRGEN WINKEL AND MAURICE OXENBURGH

Introduction

New technologies imply reduction in physical stress

At the beginning of the present decade it was estimated that about three-quarters of all operators in industrial countries performed their work in a seated position (Grandjean, 1982). It is assumed that this development will continue during the 1980s due to comprehensive computerization at most levels in working life. At the International Trade Union Conference on visual display units in 1984 it was estimated that by 1990 VDT work will be performed regularly by more than half the employees in the industrial countries (Nedzynski, 1985).

At the production line in industry it is usually most profitable to hand over physical work to machines, and operation of these is to an ever increasing extent performed by computers; the task left to the employees is mainly VDT work, or merely passive supervision (Östberg, 1977; Olsson, 1987).

During recent years office work has also been included in the computerization. In a modern computerized office the employee no longer needs to get up intermittently and walk a few steps to the loose-leaf binders to collect and file information or to exchange information with colleagues; almost all information can be reached from the VDT. Accordingly, most physical activity is eliminated.

Thus, many jobs, which traditionally were quite different and implied a large variation in physical stress during the day, are now, or will be in the near future, carried out from the same seat. The worker will be 'locked' in position in front of a terminal.

Parallel with these changes, the ergonomic design of VDT workstations is being constantly improved to reduce postural muscular activity towards zero. Many such new workstation products are, or will in the near future become, commercially available.

In summary, the introduction of VDTs implies more seated work, reduction of intermittent walking and reduction of postural load, i.e. a marked reduction in physical activity compared with the tasks which are replaced by the VDTs.

Minimizing versus optimizing physical stress

Ergonomic intervention studies are generally based on the concept that reduction in physical stress (external load) reduces physiological strain (internal reaction) and the risk of developing musculoskeletal injuries (cf. right part of the curve in Figure 6.1). More recently it has been questioned if the physical *inactivity* caused by many VDT tasks is acceptable from a physiological viewpoint (Winkel, 1981, 1987). Several investigators have reported that even when guidelines are followed and workstations are adjusted to fit individuals, a variety of musculoskeletal complaints may arise (e.g. Arndt, 1983; Oxenburgh, 1984) (cf. left part of the curve in Figure 6.1). Based on the concept illustrated in Figure 6.1, the aim of ergonomic interventions should therefore be to *optimize* rather than *minimize* physical stress.

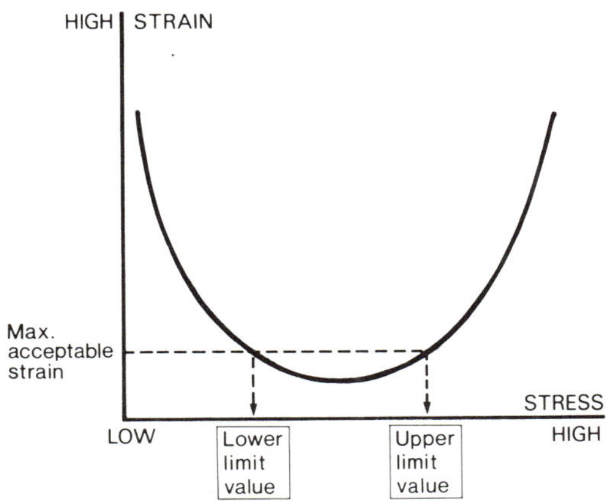

Figure 6.1. Probable relationship between stress and strain suggesting introduction of the concept 'lower limit value' in addition to 'upper limit value'. Accordingly, the aim of ergonomic interventions should be to optimize rather than reduce physical stress as far as possible. From Winkel, 1987.

Inactivity versus lack of physical variation

Bedrest studies, as well as space medicine, indicate that extreme inactivity causes detrimental effects on many physiological functions, e.g. muscles, joints, bones, circulation, electrolyte and fluid balance (for reviews, see Greenleaf, 1984; Kilbom, 1987). During bedrest, as well as space travel, three parameters have been suggested as crucial for the impairment of physiological functions (Greenleaf and Kozlowski, 1982):

1. lowered hydrostatic pressure within the cardiovascular system and body fluid compartments,
2. reduced compression on the long bones of the skeletal system, and
3. reduced energy metabolism.

In the seated position reduced levels of these parameters do not seem so marked since:

(a) the hydrostatic pressure is elevated in the lower extremities;
(b) at least some of the bones (e.g. in the spine) are continuously loaded, and
(c) the inactivity is less extreme.

Several studies suggest that it is *lack of physical variation* that explains most of the detrimental physiological effects of extreme 'inactivity' (Greenleaf and Kozlowski, 1982; Adams and Hutton, 1983; Winkel, 1985; Kraemer, 1985; Holm and Nachemson, 1983; Westgaard, 1988).

An ordinary ergonomic intervention at a VDT workstation usually implies improvements of the workstation design which minimize the physical workload. This allows longer working periods. Consequently, the exposure time (time in front of the VDT) is usually increased as much as possible to use the expensive equipment efficiently. Ergonomic benefits may therefore be lost due to efficiency goals (Winkel, 1989). It is then lack of physical variation which becomes the problem. To ensure the benefits of improved workstation design, the ergonomist has to monitor the physical variation by controlling work organization.

Pause gymnastics versus meaningful physical variation

Increased physical activity in constrained tasks is often practised by introducing pause gymnastics. This is less controversial compared with interventions to general work organization. The duration and frequency of pause gymnastics is usually low (5–10 min, once or twice a day) (Laporte, 1966) and this may be one of the reasons why it never has been possible to show any clear health effects. On the other hand, if pause gymnastics is performed more regularly during working hours it may reduce productivity. Moreover, pause gymnastics is often perceived as idling movements and this may explain why employees usually give up this kind of exercise after a while (Sundelin *et al.*, 1987). In order to obtain positive health effects and avoid reduced performance when introducing more physical variation we assume that the activity has to be evenly distributed during the day and integrated as a natural and meaningful part of the job.

Intervention at individual versus systemic level

Consequently, an intervention at systemic level is required so that the VDT work can be divided between a larger number of employees and more time is left for alternative tasks for the individual; but this may be associated with psychosocial changes, and is rarely accepted by management. It was decided that our present studies should only comprise interventions at the individual level. However, we wanted the activity to be perceived as meaningful by the operators, and therefore pause gymnastics was excluded.

How to increase physical variation at the individual level without reducing performance?

An increased physical activity may be obtained without interference with performance or changes to work organization if the workstation design allows the operator to change between sitting and standing whilst keying. Such a design was suggested early in the century (Galloway, 1919). As the importance of physical activity during recent years has become more obvious, this idea has been revived (Ishai, 1987).

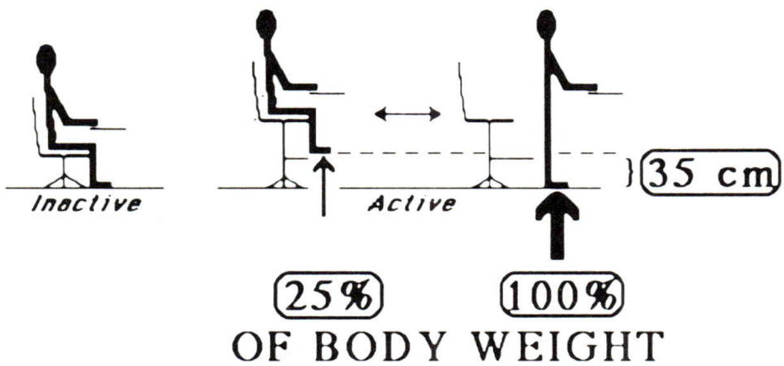

Figure 6.2. Left: A traditional workstation design. Right: An alternative workstation design which allows the operator to alternate between sitting and standing without changing the dimensions of the workstation. For further details, see text.

Standing may, however, cause increased orthostatic problems compared with the seated posture. The increased distance between heart and foot — about 35 cm (see Figure 6.2) — increases the net filtration pressure in the capillaries of the feet and may thus increase oedema formation and reduce total plasma volume (cf. Arnoldi, 1965; Hagan *et al.*, 1978; Dupuis and Rieck, 1980). In addition, the feet carry the whole body weight whilst standing but only about 25% of the body weight in the seated position (Jürgens, 1969) (Figure 6.2). These considerations should therefore be included in an intervention study.

Aim of the present study

Complaints from the lower extremities during sedentary work (Søndergård Kristensen, 1978) can be reduced by modest increase in leg activity (Winkel, 1985). Other complaints, e.g. in the shoulder–neck, which also occur at 'good' workstations (Oxenburgh, 1984) may also be reduced by a modest increase in physical variation of other body parts. It was therefore decided to carry out an intervention study at a VDT-intensive office, where musculoskeletal problems were substantial, in spite of excellent ergonomic design of the workstations. The aim was to investigate if physical activity could be increased as indicated above without reducing productivity, and thus obtain reduced discomfort and no increase in orthostatic strain reactions.

Materials and methods

Workplace

The workplace selected was in the accounts department of Televerket (Swedish Telephone Company). The workers were organized in four groups, each with a supervisor. The entire workforce was female ($n = 45$) and averaged 41·6 full-time equivalent workers during the period of the experimental work. They were occupied in entering telephone account data, via a keyboard. Most of the work was numerical with only a small percentage (less than 5%) of alphabet characters confined to certain occasions. The numerical section of the keyboard was on three rows on the right hand side of a normal QWERTY keyboard. Thus, the keyboard-entry work was performed almost entirely by the right hand. All workstations were of good quality and individually adjusted according to recommendations in standard ergonomic textbooks.

The investigation was approved by the Ethics Committee of the Medical Faculty (Karolinska Institute), Stockholm, and was divided into Study A and Study B as indicated below.

Subjects

Study A

Twenty of the operators at the department wanted to participate in the study. These went through a clinical examination of the locomotor system performed by the physiotherapist of the company. Sixteen subjects were selected on the following criteria: a minimum experience of 3 months in the present department, worked at least 75% of full-time, had no complaints which reduced their work capacity and had no circulatory problems of the lower legs and feet. Their background data are given in Table 6.1.

Table 6.1. Background data to the two investigated subject samples.

	Study A ($n = 16$)		Study B ($n = 9$)	
	Mean	Range	Mean	Range
Age (years)	32	18–60	23	18–37
Weight (kg)	58	43–72	60	51–71
Height (cm)	166	154–173	168	160–180
Blood pressure (mm Hg)				
Systolic	122	108–154	—	—
Diastolic	82	70–98	—	—
Maximum isometric torque or force			Median	Range
Dorsal flexion, right hand (Nm)	—	—	2·03	1·40–3·33
Elevation right shoulder (N)	—	—	602	498–771
Trunk extension, standing (Nm)	—	—	206	152–239
Plant. flexion, left foot (Nm)	—	—	130	97–180

Study B

Nine female operators volunteered for the study. They were selected by the same criteria as in Study A. Their background data are given in Table 6.1.

Experimental design/intervention

Study A

All the subjects were exposed to two situations:

1. an 'inactive' situation in which physical activity was reduced to a minimum likely to occur in constrained VDT work,
2. an 'active' situation in which physical activity was increased as far as practicable without reducing individual productivity.

The active situation was the experimental condition; the inactive situation was the control/placebo condition to balance the effects of a novel situation ('Hawthorne effect'; see Pennock, 1930). Both situations were presented to the subjects as suggested improvements which we wanted to test; they had new chairs, but tables and other equipment were unchanged.

In the inactive situation the Nomi 5 chair from Höganäsmöbler AB, Sweden was selected on the basis of its appearance, to give the subjects an impression of improvement (Figure 6.3A). It was presumed that this chair would cause no physiological effects different from that caused by their normal work chair. The Nomi 5, at our request, had no wheels (castors) as these might have caused an increase in physical activity and corresponding physiological effects (Winkel and Jørgensen, 1986a).

For the active situation a chair was selected that enabled the subjects to work either standing or sitting according to the principle illustrated in Figure 6.2. The Maxi

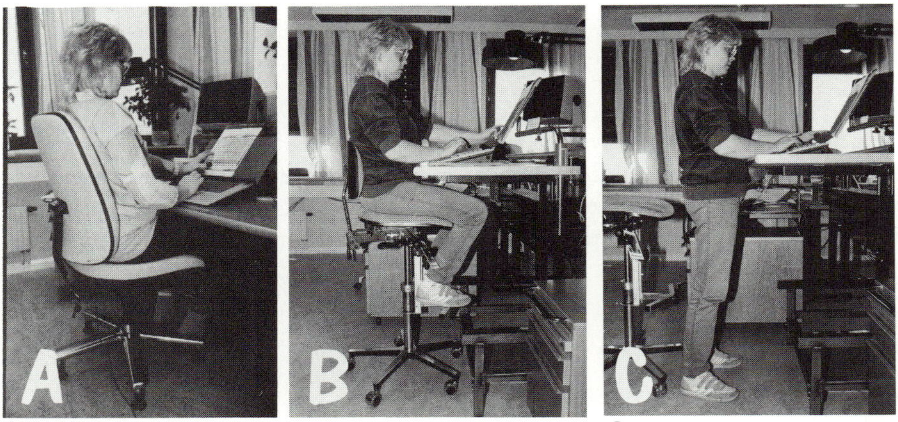

A B C

Figure 6.3. The workstations used in the 'inactive' (A) and the 'active' situations (B, C).

5/6A from RH Form AB, Sweden, met this criterion (Figure 6.3B,C). The seat could be sloped forward to make getting on and off easier, there was an adjustable bar for the feet, and wheels were also fitted. The latter had a safety device to prevent movement of the chair when sitting or while getting off.

Both workstations were adjusted so that, while wearing their usual work shoes, the subjects' right elbow was in the same position relative to the home row of keys. Seat height above the foot rest was adjusted according to popliteal height. In addition to these adjustments, further adjustments during the experimental periods were made by the subjects themselves according to their personal preferences. Arm rests were omitted as they might have interfered with arm movements; at their ordinary workstations arm rests were not fitted.

The subjects had three formal breaks during the working day: lunch, morning and afternoon coffee breaks. In the inactive situation the subjects remained seated during the working day (except for the official breaks) and data documents were delivered to their workstations. In the active situation the subjects were encouraged to change between sitting and standing at least four times per hour. They had to fetch their data documents from a table 5–10 m from their workstation.

Each subject worked for about 5 whole days in each of the two experimental work situations. On the 5th day the measurements were carried out.

Study B

The subjects were exposed to the same two situations as in Study A. However, in Study B the experimental period was reduced to 1 h for each situation and both were performed on the same day for each subject. In the 'active' situation the subjects were sitting for 9 min, changed to standing and after additional 8 min of data entry they walked 7 m to a table to fetch new data documents and 7 m back. This cycle was repeated for an hour. Thus, the variance of the investigated dependent variables could be reduced. The activity pattern (sitting/standing/fetching documents) corresponded approximately to the average measured in Study A while the subjects were keyed into the computer.

Study A and B

In Study A as well as Study B the experimental sequence was balanced between the subjects. They worked at their own pace and were free to stop or start as they required. For 2 months before the experiments each workstation type was placed in the accounts department, so that the subjects got the opportunity to become familiar with the equipment.

Measurements (Study A)

Productivity

With Union agreement, it was arranged that individual productivity data could be collected from the computers used. The computer collected data on keystrokes, time spent keying (switched into the system) and type and number of documents. As all telephone accounts are entered twice, to ensure accuracy, error rates were also obtained. They were recorded as keystrokes wrong per thousand keystrokes entered. Productivity data were collected continuously during the experiments.

Room temperature

Global temperature was measured at floor level twice a day. The average was 23°C (s.d., 0·7).

Physical activity (the independent variable)

A photoelectric cell, connected to an Event recorder (Zander and Engström, Sweden), was mounted across the chair, in the position that would be occupied by an arm rest. Thus, it was possible to measure the proportion of time spent sitting and the frequency of movements to and from their chair. The length of the working day, less time for formal breaks for lunch and coffee, could also be calculated from this recording system. The physical activity was measured for 2–3 days in each of the two experimental situations.

Discomfort ratings (DR), heart rate (HR), foot volume (FV), change in total plasma volume (ΔPV), skin temperature (T_{sk})

DR were made by the subjects on all five experimental days. The ratings were made four times each day as indicated in Figure 6.4. The subjects were asked to indicate on a

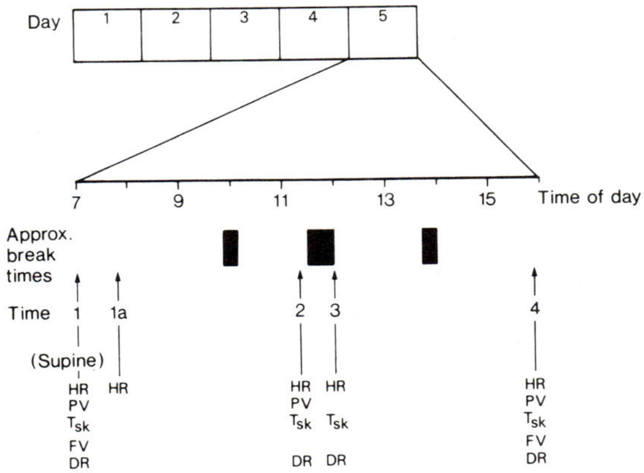

Figure 6.4. Diagram of an experimental week in Study A, showing details of the last day on which the measurements were made. Start of working day (time 1) varied between subjects from 7 to 9 a.m. and the work schedule during the day varied according to this. The measurements were: heart rate (HR), change in plasma volume (ΔPV), skin temperature (T_{sk}), foot volume (FV) and discomfort ratings (DR). Time 1a: starting work at workstation.

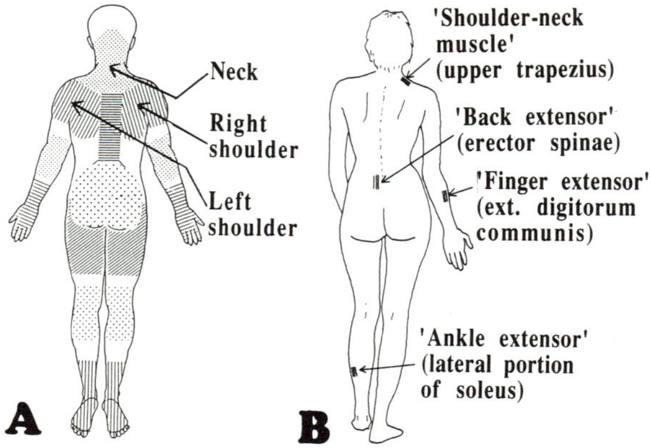

Figure 6.5. (A) The body diagram for discomfort ratings used in Study A. (B) Location of the four muscles investigated by EMG. The names in quotation marks indicate how the muscles are referred to in text.

In addition, they indicated the degree of discomfort from each of these areas on an analogue-visual (AV) 10-cm scale (Freyd, 1923). The extremes were defined as 'not at all' and 'very much = cannot work'. AV-scales for discomfort of the head, neck and shoulders always had to be filled in. The physiological parameters were only measured during the 5th experimental day (Figure 6.4). HR was stored on a Medilog (Oxford Instruments, series 4:24) during 10-min periods. Left foot volume was measured in a constant water level plethysmograph with a mean experimental error of 0·16% of the foot volume, i.e. about 1·6 ml (Winkel, 1986). Whole-body hematocrit values (Hct) were estimated using capillary blood from a pre-warmed finger. The percentage change in PV was then calculated from the Hct (van Beaumont, 1972; Winkel and Jørgensen, 1986b). For the first blood sample in the morning the subjects were supine for an average of 35 min (s.d.,5) and then blood was collected. T_{sk} was measured with thermistors fixed with adhesive tape to the dorsum of the right foot, and to the right calf, one-third of the distance up from the ankle to the knee (see Noddeland and Winkel, 1988).

Measurements (Study B)

Productivity

This was measured as number of keystrokes during the experimental hour.

Electromyography (EMG)

The details concerning instrumentation, calibration and analysis of the EMG signals were as previously described by Winkel (1985). Myoelectric activity was recorded from a right finger extensor (extensor digitorum communis), a right shoulder–neck

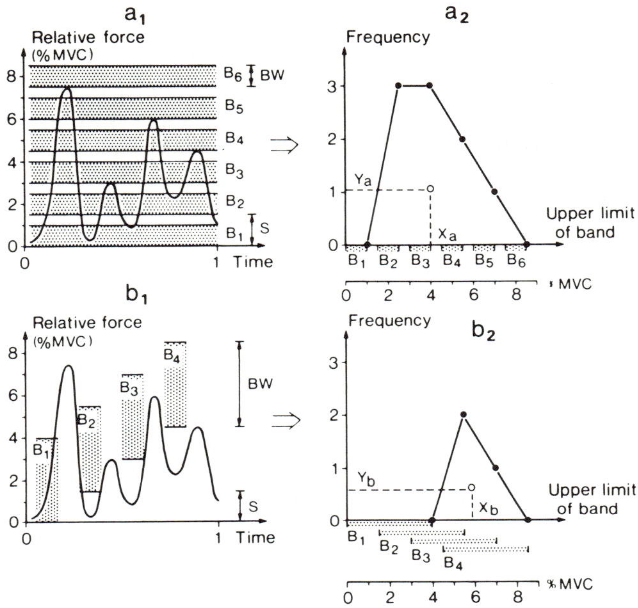

Figure 6.6. An illustration of the method used to evaluate muscular contraction frequencies. A muscular performance is analysed twice (a_1 and b_1) using different bandwidths (BW). The position of the band is raised in steps (S) from B_1 until the frequency of band crossings from below is zero. (a_2) and (b_2) indicate the corresponding frequency/force curves. Both coordinates (x and y) of the centre points (○) divide the area below the curves into halves. In (b_1) the BW is four times larger compared with the value used in (a_1). This implies an elimination of the smallest changes in tension and a corresponding displacement of the frequency/force curve: the median amplitude of the centre point increases from X_a to X_b and the median frequency decreases from Y_a to Y_b. In the present study a small value of S was used to improve the estimate of the frequency/force curve. From Winkel, 1985.

muscle (upper trapezius), a left back extensor (erector spinae, L3-level) and a left calf muscle/ankle extensor (lateral portion of soleus) (see Figure 6.5B) by means of surface electrodes. The EMG signals were recorded continuously during the two experimental hours. Before the experiments a calibration between EMG and force was obtained by a slowly increasing isometric test contraction. The amplitude probability distribution function (APDF) of the EMG signals during work was converted to an APDF of the relative force of contraction, the percentage maximal voluntary contraction (% MVC) (Hagberg, 1981; Jonsson, 1982). The APDF of the calibrated EMG signals indicates the relative time during which the muscular force has been lower than or equal to the contraction levels indicated by the APDF curve. The force of muscle contraction corresponding to the probability level (P) 0·10 in the APDF is defined as the 'static load level' (Jonsson, 1988), i.e. the muscle strain is lower than this value for 10% of the recording period. P:0·90 in the APDF is usually selected to reflect the highest level of contraction ('peak load level', cf. Hagberg, 1981); but if the investigated muscle is relaxed for prolonged periods, the probability level which corresponds to the actual peak muscular load may be higher than 0·90 (Winkel and Bendix, 1986). P:0·99 was therefore also computed and statistically analysed.

The APDF-analysis does not consider variations in contraction frequencies and important information about variation in physical activity may be lost without further EMG analyses (cf. Winkel, 1985; Winkel and Bendix, 1986). The EMG signals were therefore analysed for muscular contraction frequencies, as described by Winkel and Bendix (1984). A contraction frequency was calculated as the number of times the relative muscular force crossed an amplitude band from below (Figure 6.6). The band was moved in steps (S) of 0·25% MVC to a level corresponding to the probability level (P) 0·99 in the APDF. The centre point of the resulting frequency/force curve was then calculated. The y coordinate of this point — the median frequency — was taken as an estimate of the contraction frequencies and the x coordinate — the median amplitude — as an estimate of the force of the contractions. Each EMG recording was analysed several times, using different bandwidths (BW).

Statistical methods

The discomfort ratings and HR were analysed using a three-way ANOVA (condition, time, subjects). Due to the skew distribution of the median frequencies and amplitudes of the muscular contractions, these data were analysed using the one-sample Wilcoxon test or the Pratt test (Rahe, 1974) if zero differences occurred. The remaining data were analysed using the one-sample t-test. P-values (two-tailed) about 0·05 or lower were considered as statistically significant.

Results

The clinical examination of the 20 potential subjects indicated the following numbers of cases: 'shoulder–neck soreness' 19, 'lumbar spine soreness' 10, 'arm disorders' 9, 'leg/foot disorders' 4. Out of the 9 cases of arm disorders, 8 were assumed to be derived from the shoulder–neck region.

Measures of the physical activity in Study A are given in Table 6.2. In plain terms, the subjects rose or sat down in their chairs six times per hour during the active condition and spent 25 min/h standing to work or moving about to fetch new documents. During the inactive condition the subjects only left their chairs once for a period of less than 4 min during the whole working day (excluding the official breaks). Accordingly, Figure 6.7 shows a higher heart rate during the active situation ($p<0·01$). There was no difference in the number of keystrokes/hour (Table 6.3). The increased productivity per day in the inactive situation (Study A) was due to longer 'computer time'. The high number of keystrokes/hour in Study B (cf. Study A) is explained by the short total working time and the particular keyboard task (only numerical).

The discomfort ratings of the head, neck and shoulders are given in Figure 6.8. Only the neck region showed a significant smaller increase in discomfort rating during the active condition compared with the inactive one ($p = 0 \cdot 057$). When all four body parts indicated in Figure 6.8 were analysed together, it also appeared that this total discomfort level at the end of the working day was lowest in the active condition ($p = 0 \cdot 04$).

Both working conditions caused the same increase in foot volume (2·9%) and decrease in total PV (8%) (Figure 6.9). The loss in PV occurred only during the first half of the working day.

Table 6.2. Physical activity of the subjects during the experimental conditions in study A.

	'Inactive'		'Active'			
	Mean	s.d.	Mean	s.d.	s.e.d	*p*
Working time (h)[a]	5·6	0·8	5·4	0·6	0·2	n.s.
% Sitting[b]	99	2	59	12	3	<0·0001
Frequency (per h)[c]	0·33	0·36	6·05	2·25	0·53	<0·0001

[a] Working time, total time (h) spent at their workplace during the day excluding official breaks for lunch, morning and afternoon coffee.
[b] % Sitting, time sitting at the workstation in per cent of working time.
[c] Frequency, frequency of changes from standing to sitting or vice versa (per hour).

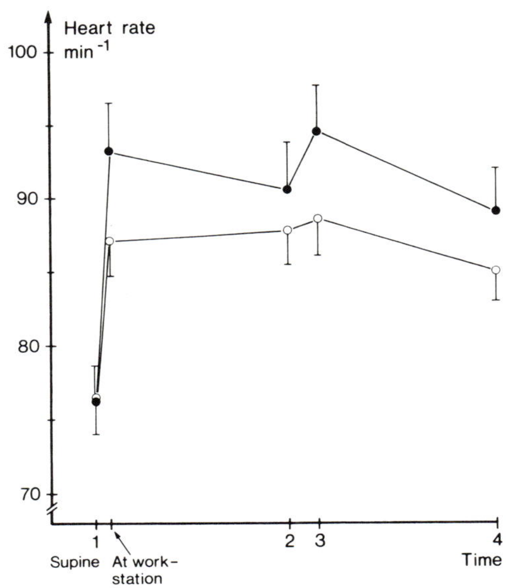

Figure 6.7. Time course of heart rate during active (•) and inactive (○) working days (Study A). The vertical bars indicate standard error of the sample means (n = *16*). *From Aronsson* et al., *1989.*

Table 6.3. Productivity per subject indicated as keystrokes/hour 'keyed into' the computer, keystrokes/day, computer time/day (= hours 'keyed into' the computer/day) and error rate (keystrokes wrong/1000 keystrokes).

	Total workforce[a]	'Inactive'		'Active'			
	Mean	Mean	s.d.	Mean	s.d.	s.e.d.	*p*
Study A							
Keystrokes/hour	8200	8900	2400	8500	2300	228	n.s.
Keystrokes/day	30900	39600	12100	30800	9900	1705	<0·0001
Computer time	3·79	4·47	0·66	3·63	0·70	0·156	<0·0001
Error rate	—	0·75	0·31	0·85	0·39	0·09	n.s.
Study B							
Keystrokes/hour		14600	2200	13700	3200	760	n.s.

[a] Total workforce, the 41·6 full-time equivalent workers in the accounts department; their mean values are calculated for those days on which the experimental work was done. s.e.d., standard error of differences.

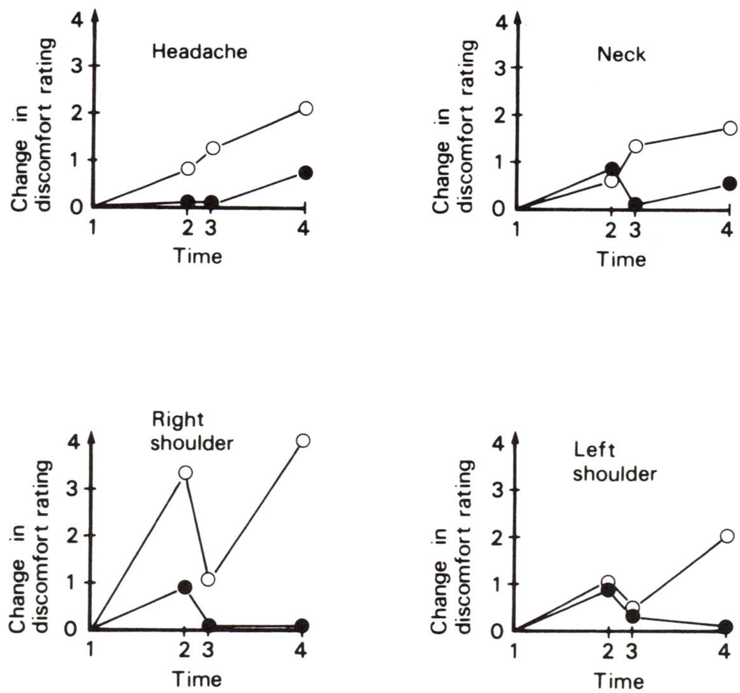

Figure 6.8. Mean changes in discomfort ratings relative to the morning values (time 1) during active (•) and inactive (○) working days (Study A, n = *16). From Aronsson* et al., *1989.*

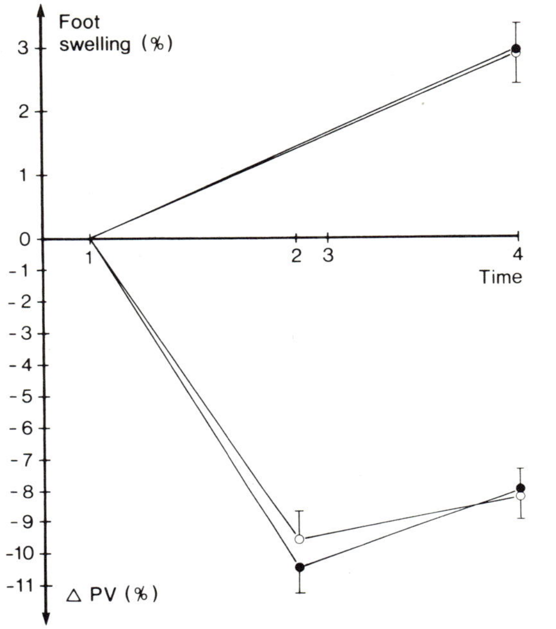

Figure 6.9. Mean changes in foot swelling and total plasma volume (ΔPV) relative to the morning values (time 1) during active (•) and inactive (○) working days (Study A). The vertical bars indicate standard error of the sample means (n = *16*). *From Aronsson* et al., *1989.*

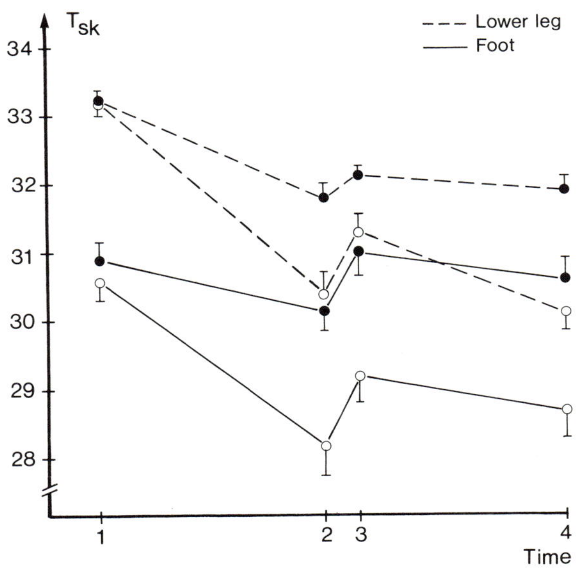

Figure 6.10. Time course of skin temperatures (T_{sk}) during active (•) and inactive (○) working days (Study A). The vertical bars indicate standard error of the sample means (n = *16*). *From Aronsson* et al., *1989.*

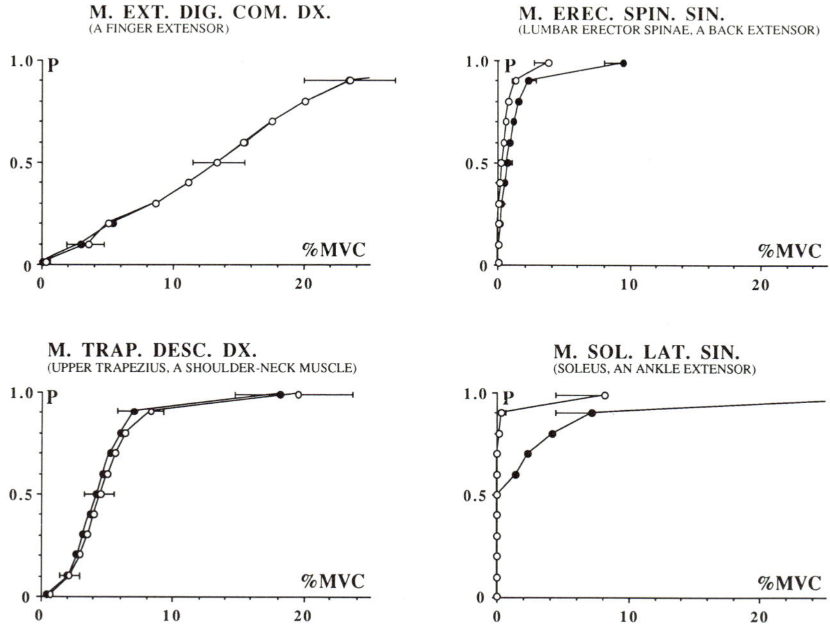

Figure 6.11. Mean APDFs of the four investigated muscles during active (•) and inactive (○) working days (Study B). The horizontal bars indicate standard error of the sample means (n = *9*).

The general pattern for the changes in T_{sk} (Figure 6.10) was that the foot as well as the lower leg were mainly cooled during the first half of the inactive working day (about 3°C). During the active day the average cooling was significantly reduced to less than 1°C ($p \leq 0{\cdot}003$).

Figure 6.11 illustrates the APDFs of the four investigated muscles. Significant differences between the active and inactive conditions occurred only for the highest load levels and only for the back extensor (P: 0·99, $p = 0{\cdot}003$) and the ankle extensor (P: 0·90, $p = 0{\cdot}03$; P: 0·99, $p < 0{\cdot}0001$).

Figure 6.12 illustrates the results of the contraction frequency analyses. For the back extensor the frequencies as well as the forces of contractions were about twice as high during the active condition as compared to the inactive ($0{\cdot}009 \leq p \leq 0{\cdot}03$). For the ankle extensor the differences were even larger ($0{\cdot}009 \leq p \leq 0{\cdot}03$). For the finger extensor and the shoulder–neck muscle the contraction frequency analyses showed no differences between the two conditions ($p \geq 0{\cdot}1$).

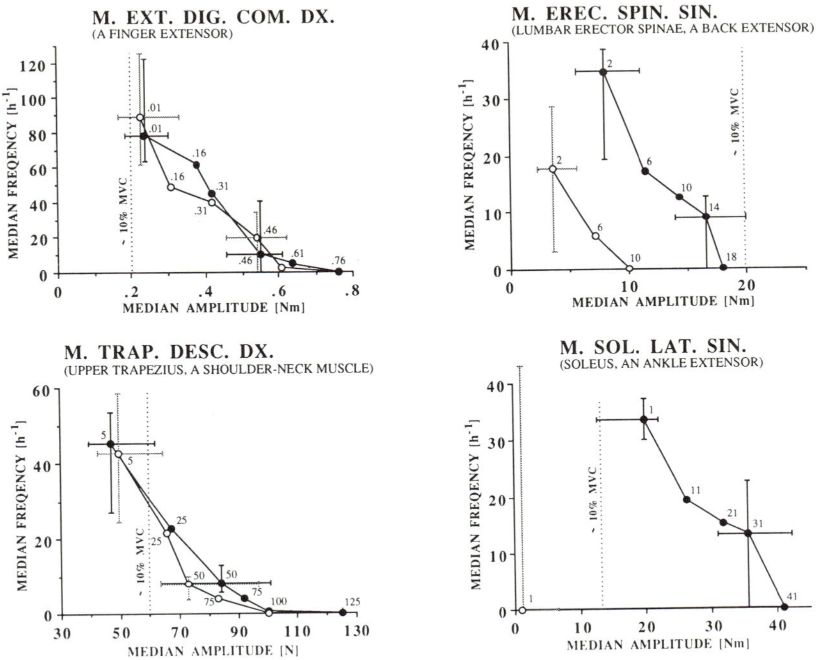

Figure 6.12. Centre points of the frequency/force curves obtained for the four investigated muscles during active (•) and inactive (○) working days (Study B). For each EMG recording different centre points were computed according to the BW (see Figure 6.6) used in the analysis. The numbers indicate the BWs in Newton metres or Newtons (upper trapezius). Each centre point is the median of nine subjects and the bars indicate the interquartile ranges between subjects. The vertical dotted lines correspond to 10% MVC for the median of the nine subjects in the corresponding test contractions (cf. Table 6.1). For further details, see text.

Discussion

Methodological considerations

In the present field study (Study A) the length of the working day was almost equal for the active and inactive conditions. However, the 'computer time' (see Table 6.3) appeared to be 50 min less during the active condition. This difference should be taken into consideration in all comparisons between the two conditions. The 50 min extra without computer work in the active condition were mainly devoted to relaxation. This was not taken into account in the design of Study B. For the whole working day the differences in leg and back muscular strain between the two conditions may therefore be larger than indicated in Figures 6.11 and 6.12.

Designation of 'neck' and 'shoulder' on body diagrams for discomfort ratings varies between investigations (cf., e.g. Corlett and Bishop, 1976). In the present study it is presumed that discomfort in the shoulder–neck muscle (upper trapezius) has been indicated as discomfort in the neck (cf. Figure 6.5A).

During recent years fundamental limitations of EMG in vocational studies have been pointed out by several authors (e.g. Christensen, 1986a, b; Westgaard, 1988). In spite of this, EMG may still provide a basis for risk estimation in showing which muscles are vulnerable in particular occupational activities (Jonsson, 1982; Winkel *et al.*, 1983a, b; Winkel and Gard, 1988).

The changes in discomfort during the working hours, as measured in Study A, are short-term and transient effects. They are, however, often considered to be early warning signals which may lead to long-term development of injuries (e.g. Hagberg, 1987).

Ergonomic and physiological considerations

Perceived discomfort

Study A shows that a change from the inactive to the active condition reduces the perceived discomfort in the neck region without reducing performance rate (keystrokes/hour) (Figure 6.8, Table 6.3). This improvement was probably due to shorter computer time (Table 6.3) and not to increased physical variation, as no differences in workload pattern in the shoulder–neck region could be shown (Figures 6.11 and 6.12). Several other investigations have shown a positive correlation between incidence of shoulder–neck complaints and duration of VDT work (Bergqvist, 1984; Oxenburgh, 1984). The active condition in Study A may also have reduced mental stress and with it strain in the shoulder–neck muscle (cf. Wærsted *et al.*, 1987); but this effect will probably not appear in a quasi-experimental field investigation as Study B.

Dynamic activity

The contraction frequencies of the shoulder–neck muscle and the finger extensor were not increased during the active condition (Figure 6.12) in spite of intermittent interruptions in keying when the subjects fetched new data documents or changed position. This is probably due to similar interruptions during the inactive condition when the subjects decided themselves about short spontaneous breaks for recovery. In the active condition the most forceful contractions comprised about 30% MVC for the finger extensor and the ankle extensor (see Figure 6.12). For the shoulder–neck muscle the corresponding value was about 20% MVC and for the back extensor less than 10% MVC. The estimate of contraction frequencies was highest for the finger extensor which is comprehensible as this muscle is a prime mover in the keying task. Of the four investigated muscles, the highest level of dynamic activity therefore seems to occur in the finger extensor. Whether this implies a beneficial or detrimental strain pattern is hard to decide. We are still not able to discriminate between monotonous and repetitive strain causing injuries and physical variation giving rise to nutritional effects preventing injuries.

THE BACK EXTENSOR

The increase in dynamic activity of the back extensor in the active condition (cf. Figure 6.12) may reflect a corresponding effect on the lumbar disc pressure (Andersson *et al.*, 1974). Furthermore, a variation in load on the spine is required to obtain an acceptable nutrition of the intervertebral discs (Adams and Hutton, 1983; Holm and Nachemson, 1983; Kraemer, 1985). Thus, for the back the active condition may imply reduced strain. Unfortunately, we do not know if the present increase in physical activity of the back extensor is sufficient to reduce the risk for development of back complaints.

THE ANKLE EXTENSOR

The pronounced increase in dynamic activity of the ankle extensor in the active condition (Figure 6.12) explains the positive effect on circulation (Figure 6.10). Although the average distance between foot and heart was increased in the active condition (see Figure 6.2), oedema formation in the foot and loss of total PV were not enhanced. This is explained by the increased leg activity stimulating lymphatic drainage and reduction in venous blood pressure (cf. Noddeland *et al.*, 1983; Noddeland and Winkel, 1988; Winkel *et al.*, 1988). A reduced venous pressure is suggested by the increase in T_{sk} (reduced venous pressure may improve circulation and thereby delivery of heat to the skin). From previous studies we also know that modest increase in leg activity performed while seated may be sufficient to reduce the perception of leg/foot complaints (Winkel, 1985). Prolonged periods in a standing posture may, however, increase leg/foot discomfort (Winkel, 1982; Buckle *et al.*, 1986; Ryan, 1989). In the present active condition none of the subjects complained of discomfort in the lower extremities, thus suggesting that the physical stress was neither too high nor too low (cf. Figure 6.1).

Static load

THE SHOULDER–NECK MUSCLE

The clinical examination of the 20 operators suggests that the main problem in data-entry work is centrally (shoulder–neck) rather than peripherally (upper extremities) located pain. This agrees with previous studies in which VDT workers have been compared with non-VDT workers (Bergqvist, 1984). In a recent review by Hagberg and Wegman (1987) it was concluded that keyboard work is associated with high risk for shoulder–neck disorders. In addition, it is often hypothesized that the causative factor is static tension (contraction) of the shoulder–neck muscles (Waris, 1979; Hagberg, 1984). Our study showed an average static load level of 2·2% MVC. This agrees fairly well with the 3·2% MVC obtained by Hagberg and Sundelin (1986) in a study of VDT word processor operators. According to Jonsson (1982) 'acceptable' static load level for long-term work should be lower than 2% MVC. However, it has recently been concluded by Westgaard (1988) that '. . . the evidence so far is that there is no acceptable limit for static muscle load . . .'. According to this, an improved ergonomic design of the workstation may never solve the crucial problem, which we assume to be lack of physical variation in the shoulder–neck region. It may therefore be necessary to integrate VDT work with other tasks where constrained postures of the neck do not occur (job enlargement/rotation).

THE FINGER EXTENSOR

The highest static load level was obtained for the finger extensor (3·3% MVC on average); but at the present workplace complaints from the right lower arm and wrist were rare compared with shoulder–neck complaints. The total activity pattern (static and dynamic) in the finger extensor may therefore be within acceptable limits (cf. above and Figure 6.1).

Concluding remarks

Our investigation emphasizes that the present challenge for ergonomic research is to characterize acceptable physical variation so undue strain on the muscles, supportive tissues and circulation is avoided. In VDT work the main problem seems to be optimization of physical stress in the shoulder–neck region (cf. Figure 6.1); but this cannot be obtained by intervention at the individual level without reducing productivity/day. Some kind of job enlargement/rotation is required. A significant organizational problem thus remains to be solved: what kind of alternative (meaningful and productive) tasks may be offered to the operators to offset the reduced time at the VDT? The answer to this question may be specific for each company, but according to our personal experience the potential for alternative tasks is greater than generally realized.

Practical implications of the present studies

1. Musculoskeletal complaints were prevalent among the operators although their ordinary (non-experimental) workstations were of good quality and individually adjusted according to recommendations in standard ergonomic textbooks. This suggests that a minimal level of physical variation is required to avoid complaints.
2. Increased physical variation may be obtained by an alternative workstation which allows the operator to change between sitting and standing without changing the dimensions of the workstation. Further increase in physical variation may be obtained if the operator has to fetch data documents and thus is encouraged to leave the VDT at intervals. These interventions do not reduce productivity (keystrokes/hour).
3. The above mentioned interventions imply an increased dynamic activity of the back extensor (erector spinae). It is unknown if this is sufficient to reduce the risk for development of chronic back complaints.

4. Although the above mentioned interventions encourage the operator to stand for about 40% of the working day, no orthostatic problems (foot swelling, loss of total PV) or discomfort in the lower extremities seem to occur. In fact, local circulation may be improved.
5. None of the above mentioned interventions seem to reduce the main problem: shoulder–neck discomfort, unless the operator at the same time reduces the total number of keystrokes during the day. Consequently, VDT work should be distributed among an increased number of employees.

Acknowledgements

We express our appreciation to all those who helped us to complete this investigation: the subjects, Nils Aava, Inga Ekman, Claes Frisk, Carina Käll, Anders Kjellberg, Nina Lindström, Nils Lundin, Lennart Möller, Björn Sköldström, Gunilla Thorstenson, Birgitta Tryggve, our co-workers Gunnar Aronsson, Francesco Gamberale and Arne Wennberg, and many others at the Swedish Telephone Company and the Swedish National Institute of Occupational Health.

This investigation was supported by grants from the Swedish Work Environment Fund (p.nr. 85-1024, 86-0419 and 86-0830), the Swedish National Institute of Occupational Health and the Australian National Commission for Occupational Health and Safety.

References

Adams, M. A. and Hutton, W. C., 1983, The effect of posture on the fluid content of lumbar intervertebral discs. *Spine*, **8**, 665–671.

Andersson, B. J. G., Örtengren, R., Nachemson, A. and Elfström, G., 1974, Lumbar disc pressure and myoelectric back muscle activity during sitting. I. Studies on an experimental chair. *Scandinavian Journal of Rehabilitation Medicine*, **6**, 104–114.

Arndt, R., 1983, Work posture and musculoskeletal problems of video display terminal operators — review and reappraisal. *American Industrial Hygiene Association Journal*, **44**, 437–446.

Arnoldi, C. C., 1965, Venous pressure in the leg of healthy human subjects at rest and during muscular exercise in the nearly erect position. *Acta Chirurgica Scandinavica*, **130**, 570–583.

Aronsson, G., Oxenburgh, M., Wennberg, A. and Winkel, J., 1989, *Optimizing Physical Activity in VDU Work. An Intervention Study*. Investigation Report 1989:24 (Stockholm: Swedish National Institute of Occupational Health).

Bergqvist, U., 1984, Video display terminals and health. A technical and medical appraisal of the state of the art. *Scandinavian Journal of Work, Environment and Health*, **10**, Suppl. 2.

Buckle, P. W., Stubbs, D. A. and Baty, D., 1986, Musculoskeletal disorders (and discomfort) and associated work factors. In: *The Ergonomics of Working Postures. Models, methods and cases*, edited by N. Corlett, J. Wilson and I. Manenica (London: Taylor and Francis), pp. 19–30.

Christensen, H., 1986a, Muscle activity and fatigue in the shoulder muscles during repetitive work. *European Journal of Applied Physiology and Occupational Physiology*, **54**, 596–601.

Christensen, H. 1986b, Muscle activity and fatigue in the shoulder muscles of assembly plant employees. *Scandinavian Journal of Work, Environment and Health*, **12**, 582–587.

Corlett, E. N. and Bishop, R. P., 1976, A technique for assessing postural discomfort. *Ergonomics*, **19**, 175–182.

Dupuis, H. and Rieck, A., 1980, Zur Belastung und Beanspruchung des Verkaufspersonals durch lange Stehtätigkeit. *Zietschrift für Arbeitswissenschaft*, **34**, 56–60.

Freyd, M., 1923, The graphic rating scale. *Journal of Educational Psychology*, **14**, 83–102.

Galloway, L., 1919, *Office Management: Its Principles and Practice* pp. 191–193 (New York: Ronald Press).

Grandjean, E., 1982, *Fitting the Task to the Man* (London: Taylor and Francis).

Greenleaf, J. E., 1984, Physiological responses to prolonged bed rest and fluid immersion in humans. *Journal of Applied Physiology: Respiration, Environment, Exercise Physiology*, **57**, 619–633.

Greenleaf, J. E. and Kozlowski, S., 1982, Physiological consequences of reduced physical activity during bed rest. In: *Exercise and Sport Sciences Reviews*, Vol. **10**, edited by R. L. Terjung, pp. 84–119 (Philadelphia, PA: Franklin Institute Press).

Hagan, R. D., Diaz, F. J. and Horvath, S. M., 1978, Plasma volume changes with movement to supine and standing positions. *Journal of Applied Physiology: Respiratory, Environmental and Exercise Physiology*, **45**, 414–418.

Hagberg, M., 1981, On evaluation of local muscular load and fatigue by electromyography. Arbete ach Hälsa 1981:24. Arbetarskyddsverket, Stockholm (Thesis).

Hagberg, M., 1984, Occupational musculoskeletal stress and disorders of the neck and shoulder: a review of possible pathophysiology. *International Archives of Occupational and Environmental Health*, **53**, 269–278.

Hagberg, M., 1987, *Occupational Shoulder and Neck Disorders*. The Swedish Work Environment Fund, No. 40-19-17 (Tibatryck City, Sweden).

Hagberg, M. and Sundelin, G., 1986, Discomfort and load on the upper trapezius muscle when operating a wordprocessor. *Ergonomics*, **29**, 1637–1645.

Hagberg, M. and Wegman, D. H., 1987, Prevalence rates and odds ratios of shoulder–neck diseases in different occupational groups. *British Journal of Industrial Medicine*, **44**, 602–610.

Holm, S. and Nachemson, A., 1983, Variations in the nutrition of the canine intervertebral disc induced by motion. *Spine*, **8**, 866–874.

Ishai, E., 1987, The effects of VDU on the interior design of offices. In: *Work with Display Units*, edited by B. Knave and P.-G. Widebäck, pp. 320–328 (Amsterdam: Elsevier/North-Holland).

Jonsson, B., 1982, Measurement and evaluation of local muscular strain in the shoulder during constrained work. *Journal of Human Ergology (Tokyo)*, **11**, 73–88.

Jonsson, B., 1988, The static load component in muscle work. *European Journal of Applied Physiology and Occupational Physiology*, **57**, 305–310.

Jürgens, H. W., 1969, Verteilung des Körperdrucks auf Sitzfläche und Rückenlehne als Problem der Industrieanthropologie. *Ergonomics*, **12**, 198–205.

Kilbom, Å., 1987, Short- and long-term effects of extreme physical inactivity — a review. In: *Work with Display Units*, edited by B. Knave and P.-G. Widebäck, pp. 219–228 (Amsterdam: Elsevier/North-Holland).

Kraemer, J., 1985, Dynamic characteristics of the vertebral column, effects of prolonged loading. *Ergonomics*, **28**, 95–97.

Laporte, W., 1966, The influence of a gymnastic pause upon recovery following post office work. *Ergonomics*, **9**, 501–506.

Nedzynski, S., 1985, International trade union guidelines for visual display units (Internationella fackliga riktlinjer för bildskärmsterminaler) (Geneva: FIET). In Swedish.

Noddeland, H. and Winkel, J., 1988, Effects of leg activity and ambient barometric pressure on foot swelling and lower-limb skin temperature during 8 h of sitting. *European Journal of Applied Physiology and Occupational Physiology*, **57**, 409–414.

Noddeland, H., Ingemansen, R., Reed, R. K. and Aukland, K., 1983, A telemetric technique for studies of venous pressure in the human leg during different positions and activities. *Clinical Physiology*, **3**, 573–576.

Olsson, G., 1987, Technological progress and the work and working conditions of process operators, University of Luleå, Sweden, 1987:58 D. (Thesis). In Swedish with an abstract in English.

Östberg, O., 1977, *Från Arbetare Till Operatör*. Technical Report 1977:50 T (Sweden: University of Luleå). In Swedish.

Oxenburgh, M., 1984, Musculoskeletal injuries occurring in word processor operators. *Proceedings of the 21st Annual Conference of the Ergonomics Society of Australia and New Zealand*, Sydney, pp. 137–143.

Pennock, G. A., 1930, Industrial research at Hawthorne. An experimental investigation of rest periods, working conditions and other influences. *Personnel Journal*, **8**, 296–313.

Rahe, A. J., 1974, Tables of critical values for the Pratt matched pair signed rank statistic. *Journal of the American Statistical Association*, **69**, 368–373.

Ryan, G. A., 1989, The prevalence of musculo-skeletal symptoms in supermarket workers. *Ergonomics*, **32**, 359–371.

Søndergård Kristensen, T., 1978, *Women's Work Environment* (*Kvinders arbejdsmiljø*). (Copenhagen: Fagbevægelsens Forskningsråd, Fremad/LO). In Danish.

Sundelin, G., Hagberg, M. and Kolmodin-Hedman, B., 1987, No benefits of teaching ergonomy among workers? *XXII International Congress on Occupational*

Health Work for Health, 27 September–2 October 1987, Sydney, Australia. Abstracts, p. 214.

van Beaumont, W., 1972, Evaluation of hemoconcentration from hematocrit measurements. *Journal of Applied Physiology*, **31**, 712–713.

Waris, P., 1979, Occupational cervicobrachial syndromes. *Scandinavian Journal of Work, Environment and Health*, **6**, Suppl. 3, 3–13.

Wærsted, M., Bjørklund, R. and Westgaard, R. H., 1987, Generation of muscle tension related to a demand of continuing attention. In: *Work with Display Units*, edited by B. Knave and P.-G. Widebäck, pp. 288–293 (Amsterdam: Elsevier/North-Holland).

Westgaard, R. H., 1988, Measurement and evaluation of postural load in occupational work situations. *European Journal of Applied Physiology and Occupational Physiology*, **57**, 291–304.

Winkel, J., 1981, Swelling of the lower leg in sedentary work — a pilot study. *Journal of Human Ergology (Tokyo)*, **10**, 139–149.

Winkel, J., 1982, An ergonomic evaluation of foot complaints among waiters as a basis for job design. In: *IEA '82, The 8th Congress of the International Ergonomics Association*, edited by K. Noro, pp. 630–631 (Japan: Inter Group).

Winkel, J., 1985, On foot swelling during prolonged sedentary work and the significance of leg activity. Arbete och Hälsa 1985:35. Arbetarskyddsverket, Stockholm (Thesis).

Winkel, J., 1986, An accurate plethysmograph for measurement of changes in the volume of the human foot. *Ergonomics*, **29**, 801–805.

Winkel, J., 1987, On the significance of physical activity in sedentary work. In: *Work with Display Units*, edited by B. Knave and P.-G. Widebäck, pp. 229–236 (Amsterdam: Elsevier/North-Holland).

Winkel, J., 1989, On the increase of occupational musculo-skeletal injuries. *Nordisk Medicin*, **104**, 324–327. In Swedish with an abstract in English.

Winkel, J. and Bendix, T., 1984, A method for electromyographic analysis of muscular contraction frequencies. *European Journal of Applied Physiology and Occupational Physiology*, **53**. 112–117.

Winkel, J. and Bendix, T., 1986, Muscular performance during seated work evaluated by two different EMG methods. *European Journal of Applied Physiology and Occupational Physiology*, **55**, 167–173.

Winkel, J. and Gard, G., 1988, An EMG-study of work methods and equipment in crane coupling as a basis for job redesign. *Applied Ergonomics*, **19**, 178–184.

Winkel, J. and Jørgensen, K., 1986a, Evaluation of foot swelling and lower-limb temperatures in relation to leg activity during long-term seated office work. *Ergonomics*, **29**, 313–328.

Winkel, J. and Jørgensen, K., 1986b, Swelling of the foot, its vascular volume and systemic hemoconcentration during long-term constrained sitting. *European Journal of Applied Physiology and Occupational Physiology*, **55**, 162–166.

Winkel, J., Ekblom, B., Hagberg, M. and Jonsson, B., 1983a, The working environment of cleaners. Evaluation of physical strain in mopping and swabbing as a basis for job redesign. In: *Ergonomics of Workstation Design*, edited by T. O. Kvålseth, pp. 35–44 (London: Butterworth).

Winkel, J., Ekblom, B. and Tillberg, B., 1983b, Ergonomic and medical factors in shoulder/arm pain among cabin attendants as a basis for job redesign. In: *Biomechanics VIII-A*, edited by H. Matsui and K. Kobayashi, pp. 567–573 (IL: Human Kinetics Publishers).

Winkel, J., Jørgensen, K. and Noddeland, H., 1988, Significance of ambient temperature for foot swelling and oedema-preventing effect of modest leg activity while seated. *Proceedings of 10th Congress of the International Ergonomics Association*, Sydney, 1–5 August, Vol. 1, pp. 140–142.

Chapter 7

Effect Of Exercise Breaks On Musculoskeletal Strain Among Data-Entry Operators: A Case Study

DAVID A. THOMPSON

Introduction

Approximately 60 million workers now use video display terminals (VDTs) at work (Hales, 1989); a large proportion of these are involved in data-entry work.

A recent measure of the impact of data-entry work on the operators' health comes from the Third Annual VDT Health Survey conducted by the Data Entry Management Association (Bodek, 1990). This survey sampled 514 operators and clerical employees from 31 data processing organizations. Data-entry personnel reported the following incidence of health problems:

(a) painful neck and shoulders, 64%
(b) back pain, 52%
(c) pain in arms and legs, 28%
(d) swollen muscles and joints, 12%.

Clearly, these results (and similar ones reported extensively by others) indicate a significant health problem.

This chapter describes the results of data-entry job redesign in a large northern California public utility. Intervention into the jobs primarily involved the introduction of paid exercise breaks for the data-entry operators. The purpose of this job change was to relieve musculoskeletal strain and its effect on the health and well-being of the operators. The results of the exercise break programme were followed for approximately 1 year after the job changes were made.

The effectiveness of excercise in relieving musculoskeletal strain has been studied in the laboratory (Lee and Humphrey, 1985; Lee and Waikar, 1986), but no field studies have been found that evaluated actual jobs under existing work stress. Lee and Humphrey (1985) found that exercise reduces the strain of the neck and trapezius muscles caused by rigid work postures. Both studies found that exercise can 'contribute to the reduction of muscle fatigue as much as ergonomic features of the (well designed) work station.'

Data-entry work is characterized by postural rigidity accompanied primarily by rapid motions of one hand over a limited range. The static posture is required in order to touch enter the data (mostly numeric) without looking at the keyboard. In order to accomplish a 'blind' touch entry, the shoulder girdle, arms and wrist must be held sufficiently rigid over time to allow repeated finger positioning from memory with an accuracy within $\pm \frac{1}{4}''$ at the finger tip. This rigid posture must be maintained for 6–7 h per day, with limited interruption, for most entry jobs.

In the company studied (a large California public utility), operators were already taking their allotted 15-min rest breaks at mid-morning and mid-afternoon. However, it appeared that these breaks did little to alleviate physical strain, since they typically involved sitting with poor posture on an overstuffed sofa in the lounge drinking coffee and smoking.

Exercise breaks, on the other hand, would stretch the affected muscle groups to stimulate blood circulation, reducing the lactic acid concentration in these muscle groups caused by the postural rigidity. The exercise would also increase breathing rate to assist in the oxygenation of lactic acid back into glycogen to fuel new muscular activity.

The introduction of exercise breaks was discussed with a labour–management task force consisting of union officials, managers and a company industrial engineer. Their concurrence and co-operation was elicited and obtained for the project. The data-entry operators studied worked in a Cash Management section, processing cheques and invoices, and were responsible for accurately accounting for several billion dollars of paper a year.

Modest workplace design changes were introduced at about the same time that the exercise break programme was started. These changes involved reducing some reach distances and changing the method of opening the envelopes containing incoming cheques. However, the workplace could not be made 'ergonomically correct' because the machinery used by the operator could not be modified in height, size or configuration without violating the machine manufacturer's warranty. Thus, the workplace design changes made were only of minor consequence and were not seen as having any significant impact on the productivity or musculoskeletal stress.[1] The chairs used were ergonomically well designed and were not changed as part of the study.

Method

The exercises to be included in the new programme were developed in the following manner. An exercise physiologist was shown the data-entry jobs and asked to construct a battery of exercises intended to counteract the musculoskeletal stresses that were present in the job.

This was done and, following review by the labour–management task force, the battery was taught to all 85 of the data-entry operators. The exercises were intended primarily to relieve cervicobrachial postural strain, but also included arm, wrist and

[1] Minor changes of this sort are a common occurrence in studies which focus on actual jobs, to clean up some obvious inefficiencies that should not be involved in the analysis.

lower leg manipulation. They were general in nature, involving stretching and manipulating the arms, neck, back and legs. They resembled those illustrated in Verbatim (1984), OCLI (1987), Pearce (1984), Sauter *et al.* (1986) and Tasker and Gore (1985).

The 8-h day was restructured to include the exercise breaks, (two 5-min periods per day) in addition to the normal rest breaks (which were continued as usual). A required work period of 6·5 hours of 'logged-on' time remained as before.[2] The operators were encouraged to:

1. take the full 5-min time twice a day, preferably mid-morning and mid-afternoon,
2. utilize additional time from their normal rest breaks to exercise instead of sitting,
3. utilize personal time before and after work to do the recommended exercises on their own time, at work or at home,
4. emphasize those particular exercises that seemed to relax their muscle tension the most, and which 'felt good' in terms of relieving discomfort,
5. if, during the normal workday — outside of programmed break times — muscular discomfort became evident, they should take a quick 'mini-break' of 20–30 s to stretch, flex and shake the affected muscle groups.

The exercise programme was introduced to the operators with an explanation of the therapeutic value of exercise, its function in increasing blood circulation through the muscle groups that cause discomfort, and how the exercises should be performed. The exercises were simple, easily remembered, safe and could all be done without embarrassment while wearing a dress. Individual 'variations on the theme' of the exercise were encouraged if an operator preferred to alter exercise method or timing.

All of the operators appeared motivated to engage in the exercise break programme; none objected to it. Undoubtedly, the involvement of the labour–management task force in the programme gave useful support and encouragement to its successful implementation.

Results

Health and health costs

There were direct programme pay-offs for both the operators and for management. In informal follow-up interviews the operators reported reduced discomfort at work and generally improved physical conditions outside of work.[3] In the 2 years before the introduction of the exercise programme there were, at any one time, 7–12 outstanding workman's compensation claims on behalf of injured operators claiming CTS, tendonitis, tenosynovitis or musculoskeletal pain (usually cervicobrachial).

---

[2] The new 10-min exercise allowance replaced a previous clean-up allowance which was no longer needed. Clean-up was now done by janitors.

[3] A formal questionnaire was not utilized to assess specific changes in body part discomfort.

(The exact count of claims related to data entry was complicated by accusations of malingering and under-reporting.) In the year following the introduction of the exercise break programme, there were no new claims and all of the operators out on disability leave returned to work in Cash Management, found new employment elsewhere in the company, or left the company for personal reasons. This illustrates the relative reduction in discomfort level due to the therapeutic effect of relevant, timely exercises.

There was also a direct pay-off to the company's workman's compensation costs, since a reduced caseload means lower insurance premiums in the long run (or lower health costs if self-insured). Estimates of these savings were not available, but are generally substantial at the injury level formerly experienced by the company.

Productivity

From a department manager's perspective, there was an immediate dramatic improvement in productivity. The average number of items processed per operator increased by about 25% following introduction of the exercise break programme (between May and August of 1985) compared with the same period of 1984.[4]

The relative difference in 'regular mail' (standard customer payments received) processing rates during the first year of the new programme (1985–6) compared with a year earlier is shown in Figure 7.1. The relative difference in 'full pay agency mail' (customer payments made at a retail outlet and sent to the centre) for the years 1984–5 and 1985–6 is shown in Figure 7.2. Figures 7.1 and 7.2 start in May for each year, since this was the first effective month in 1985 of the exercise break programme.

Using the same time frame, Figure 7.3 shows the number of data-entry operators working in the Cash Management centre. Although there is a seasonal trend in the data, no significant differences existed between the year in which the exercise break programme was implemented and the prior year.

However, the effect of the increased productivity of the centre during 1985–6 is reflected in the significant reduction in the number of hours of overtime worked that year (at a 50% labour premium), as shown in Figure 7.4. During 1984–5, the reduced number of data-entry operators working at the centre in the autumn and winter (see Figure 7.3) required increased overtime hours in order to meet the centre's throughput needs. However, this overtime increase was not necessary in 1985–6 because of the increased output per person as reflected in Figures 7.1 and 7.2

Another financial advantage to the utility was the drastic reduction in customer cheques waiting to be processed and sent to the bank for deposit in the utility's account. Cheques earn daily interest after banking, and although the interest rate may be only around 0.02%/day (6%/year), substantial interest earnings are generated at the cash flow level of large companies. For example, Figure 7.5 shows the value of the differing backlog of cheques delayed in processing within the centre between May

[4] Please note that the effects of the exercise break programme by itself is reflected only between May and August 1985 since another pay effect (explained later) became effective in August.

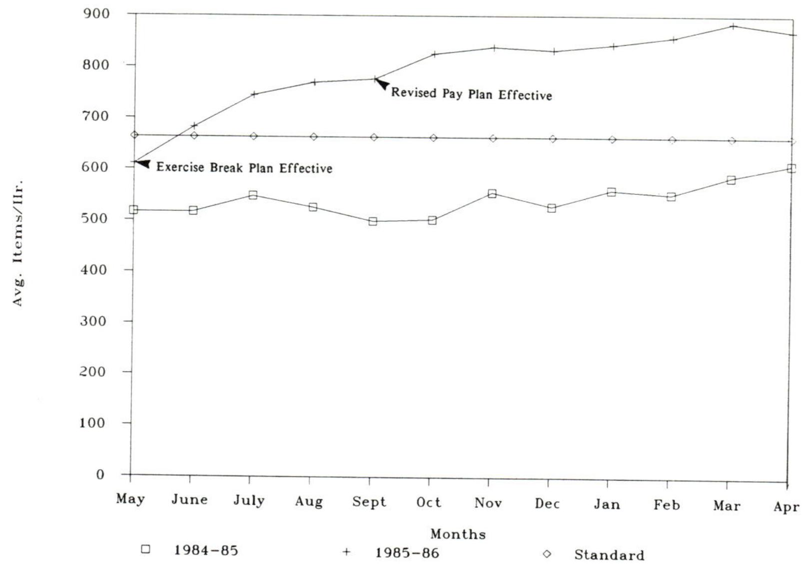

Figure 7.1 Average number of items of regular mail processed per hour in 1984–5 and 1985–6.

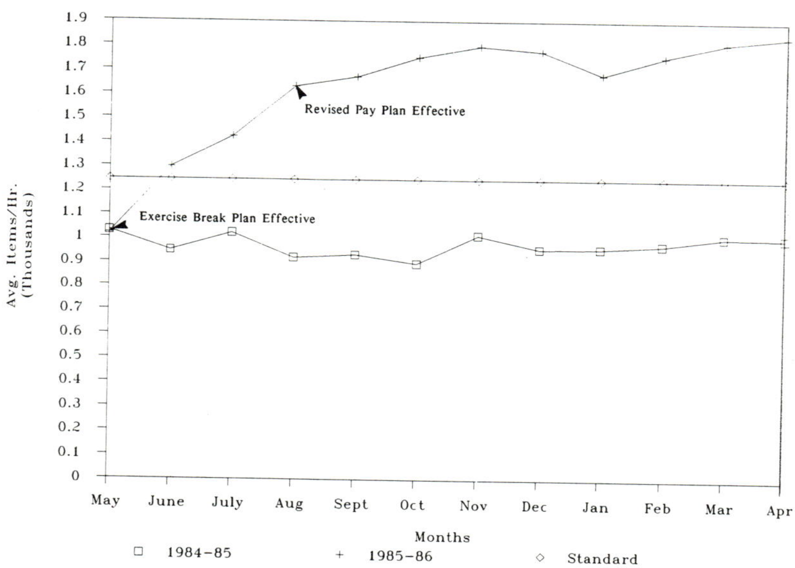

Figure 7.2 Average number of items of full pay agency mail processed per hour in 1984–5 and 1985–6.

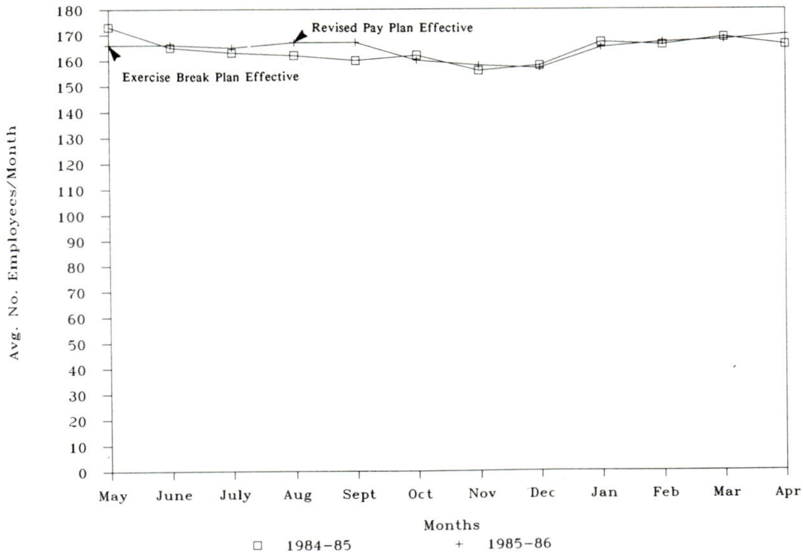

Figure 7.3 Cash management centre employees for 1984–5 and 1985–6.

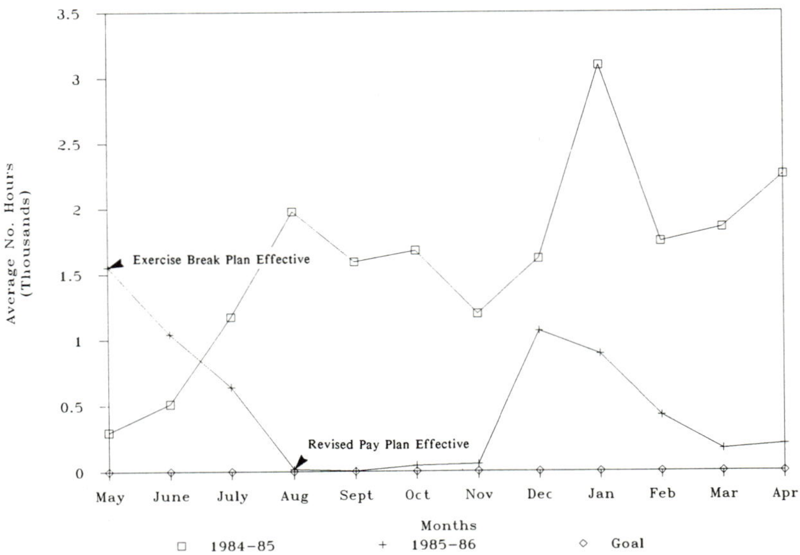

Figure 7.4 Overtime hours per month for 1984–5 and 1985–6.

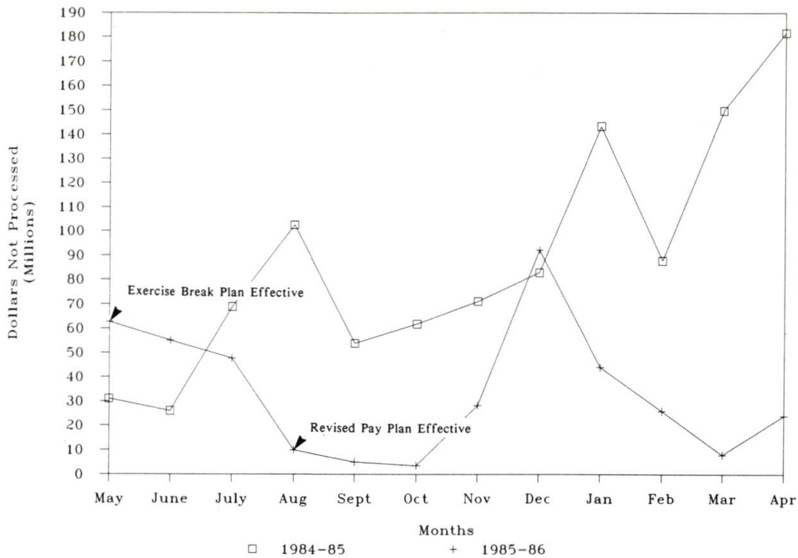

Figure 7.5 The value of the backlog of cheques delayed in processing for 1984–5 and 1985–6.

1984 and April 1986. During the period May–July 1984, the first 3 months of influence by the exercise break programme, the value of cheques delayed at least 1 day averaged US $42 million per month. During the comparable period of 1985, the value of cheques delayed at least 1 day averaged US $55·1 million per month. This alone appears not to favour the effect of exercise breaks. However, the first full year of the exercise break programme had a monthly average of cheques delayed of only US $33·88 million, whereas the previous year averaged US $88·46 million each month. This decrease of US $54·58 million following the introduction of exercises represents increased interest earnings of US $10 916 per month, at 6% per annum, or additional annual savings of US $130 992.

The above analysis requires some additional explanation. The US $130 992 interest savings during the first full year of the exercise break programme may be attributed only partially to the effect of exercise, since an incentive pay programme became effective 3 months after the start of the exercise programme. The proportion of the interest savings attributable to exercise and that contribution due to the new pay programme is discussed in the next section. It seems clear, however, that the effect of the exercise break programme in realizing these savings was significant.

The 3-month delay in beginning to realize these interest savings following the introduction of exercise breaks requires some explanation. Management was surprised at the dramatic increase in operator output early in the exercise programme (see Figure 7.1) and had not provided sufficient late night staff to have the greater number of cheques processed trucked to the bank to benefit from the extra night's interest during the May–July 1985 period. Consequently, the interest savings could have been even greater had this productivity increase been anticipated. As noted before, however, this increased productivity did show up as decreased overtime hours following introduction of the exercise break programme (see Figure 7.4).

Changes in the pay system

As discussed earlier, an incentive pay system was introduced into Cash Management 2–3 months following the start-up of the exercise break programme. Prior to 1985, the data-entry operators were paid on an hourly basis, regardless of the quality or quantity of their output. Hence there was no incentive to do one's best, and the more competent, motivated employees were even criticized by their slower colleagues for 'showing off' and for 'busting the curve'.

Following lengthy discussions with the labour–management task force, a straight piece-rate work incentive pay plan was designed and adapted. Under this plan, accurate time standards were set by industrial engineering consultants. Data-entry operators would be paid an hourly rate plus a flat rate per document processed in excess of the standard rate set by the consultants. The net result of this was that operators would be paid in direct proportion to their output, and the direct processing cost to the utility would be a constant amount per document. The indirect costs ('overhead') would be inversely proportional to output per operator, so that increased operator productivity reduced indirect costs (and therefore total costs).

The exercise break programme was effectively introduced in May 1985 as described above and the changes in productivity during the next 2–3 months can be attributed solely to its impact. However, implementation of the incentive pay plan started in July and was fully effected within the next month or two.

Thus, by August 1985 data-entry production, monthly overtime averages, and the unbanked mail backlog reflected *both* the exercise break programme and the incentive wage payment plan; these effects are also illustrated in Figures 7.1–7.5. Separating out these two effects was not possible under the actual operating conditions of the study. In fact, the two effects interacted with each other, in that the incentive wage payment plan motivated (stressed) the operators to produce at higher rates, causing more physical strain; and the exercise break programme relieved physical strain allowing higher output rates. (This interaction had been anticipated and was the primary purpose for introducing the exercise break programme.)

In discussing with the data-entry operators the relative contribution of the exercise break programme and the incentive wage payment programme to increased production, the consensus of opinion was that the exercise break programme was at least as important as the incentive wage payment programme, and that the former resulted in a much more stress-free introduction and implementation of the latter. The operators also enjoyed earning the incentive wage bonuses (typically 25–45% by the end of 1985) with reduced physical discomfort.

Discussion

Exercise programmes in industrial companies are not new. The attempt here, however, was to introduce an exercise programme that was both relevant and timely, i.e. the exercises that were performed were designed to be directly *relevant* to the particular job done, rather than have all employees do the same general exercises regardless of their jobs (as is done in many Asian companies).

In addition, the attempt was to have the exercise performed in a *timely* manner, as the musculoskeletal discomfort they were intended to relieve built up. Thus, the exercise breaks were scheduled during the work day with further encouragement to take additional mini-breaks or frequent micro-breaks. Clearly, the breaks are most effective when performed as the musculoskeletal strain begins to build up rather than after it has created physical tension and discomfort. That is, while exercise before or after work may be effective in relieving musculoskeletal strain, the same amount of exercise interspersed throughout the day is seen as much more effective.

Another, very powerful, aspect of the exercise break programme is that it puts preventive and remedial tools in the hands of those who need them. Rather than sending employees experiencing work-related musculoskeletal pain to the company medical clinic, they are given exercise techniques to prevent the pain or to ameliorate it if it should develop.

Should an operator complain to a co-worker or supervisor about musculoskeletal discomfort or pain, an appropriate, gentle rejoinder might appropriately be, 'Have you been doing your exercises regularly?' If so, and the pain persists, then the company medical clinic is the best next stop.

It was found that the programme began to deteriorate a bit for lack of management support within a year or so. Innovative programmes of this sort will not continue just because they seem like a good idea or even when found effective by many people. Old work habits change slowly, and the continuing pressures of processing the paperwork to meet quantity and quality goals and deadlines (the primary purpose of the centre) will force out other competitors for operator time if allowed to. This must be guarded against with continuing vigilance in order to reap the clear rewards of the programme.

Summary

It was found, as a result of this company's experience, that exercise breaks included in the workday's schedule can effectively reduce the musculoskeletal strain and discomfort caused by the postural rigidity of data-entry operators, thereby reducing lost time due to injury and increasing operator performance.

These breaks are seen as more effective the more they are interspersed as 'mini-breaks' or 'micro-breaks' throughout the day, although no specific evidence of this was available from the study.

The specific choice of exercises in a given situation should be made so as to accomplish non-painful stretching and manipulation of the body parts and muscle groups held most rigid during work. They should of course be safe and in good taste for those wearing dresses and street shoes, and should be explained and encouraged through frequent reinforcement to have any lasting effect on the office life style.

References

Bodek, M., 1990, VDT Health Survey, *DEMA NEWS*, p. 1 (Norwalk, CT: Data Entry Management Association).

Hales, S. W., 1989, 'Safe' VDTs. *Occupational Health and Safety*, **58** (7), 21.

Lee, K. S. and Humphrey, L., 1985, Comparison of the relative effectiveness of ergonomic design and physical exercise. In *Ergonomics International 85*, edited by Brown, I. D. *et al.*, pp. 556–558 (London: Taylor and Francis).

Lee, K. and Waikar, A., 1986, Reduction of physical stress experienced by VDT operators. *Work With Display Units Conference Proceedings*, Stockholm, 12–15 May, pp. 507–510 (Stockholm: National Board of Occupational Safety and Health).

OCLI, 1987, *Take an Exercise Break*... Santa Rosa, CA: Glare/Guard Products. 8 pp.

Pearce, R. B., 1984, *Work With VDT's: The Human Interface*, Daly City, CA: Krames Communications. 8 pp.

Sauter, S. L., Chapman, L. J. and Knutson, S. J., 1984, *Improving VDT Work*, Wisconsin: Department of Administration.

Tasker, D. and Gore, A., 1985, *Avoid RSI, Apply the Rules*. Sydney: Metropolitan Daily Newspaper Publishers and the Australian Journalists' Association.

Verbatim Inc., 1984, *Tone Up at the Terminals*. Sunnyvale, CA:

PART II
Organizational approaches and issues in the redesign of VDT work

Chapter 8

Perspectives on Ergonomics Issues in a VDT Office

OLOV ÖSTBERG[1] AND YNGVE K. O. HÖGBERG[2]

Tripartite framework

'Codetermination,' 'joint efforts', 'adherence to agreements' and 'mutual respect' are some of the keywords used by Swedish labour market parties when referring to their approach to working life issues. The keywords depict a fair amount of agenda overlap between the parties, and a sense of their being part of a win/win game. Typical comments from outside observers are as follows:

> A major contribution which Scandinavian unions and researchers have made is to address the work environment in holistic terms. Contrary to the tendency in the US to limit discussion to traditional job safety and health factors or physical and chemical hazards, the Scandinavians have understood that the psychological hazards of job stress and the nature of work relations are integrally tied to one another. As a result, the introduction of new technology into the workplace is approached as part of a larger set of worker and union rights to constantly improve the workplace. (Deutsch, 1982)

> [The Senior Vice President of the Swedish Employers' Confederation] doubted whether one conventional production line had been built in Sweden since the advent of Volvo's Kalmar assembly plant in 1974. The plant set out to design jobs in auto assembly which Swedes would wish to work in, and set forth a social specification for work before designing the technological process. . . .
>
> The Swedish approach attempts to match work to Swedish people and their conditions, the conventional approach matches the work conditions to those of the competitors, and uses the people as operators. So ergonomically speaking the Swedes are right and convention is wrong. So why not [in the UK] do what they do? (Corlett, 1987)

> Sweden is a country which could serve as a model with respect to safety and health issues and answers. Regulations development and enforcement, for example, are

[1] The Swedish Agency for Administrative Development, Box 34107, 10026 Stockholm, Sweden.

[2] Marketing Department, Swedish Telecom, 123 86 Farsta, Sweden.

characterized by a cooperative, rather than adversarial, relationship among workers, management and government regulatory agencies. The regulation development process differs from the one in the United States, in that proposed regulations are subject to a remiss process, in which deliberations resulting in a consensus occur. (Gardner-Bonneau, 1987)

Within Sweden the approach commented upon in the above quotations is proudly referred to as 'the Swedish model'. It is a tripartite approach in which trade unions, employer organizations and government regulators work together to create a nationwide framework.

This chapter is an elaboration on the manifestation of this model when it comes to ergonomics issues. The context is the use of visual display terminals (VDTs) in telephone directory services of the Swedish Telecommunications Administration (STA). This is an area of continuous negotiation, irrespective of the obvious similarities in the views held by STA local management (Törnqvist and Hermansson, 1983) and local trade unions (Westman, 1983).

The bottom line of ergonomics

The Codetermination Act and the Working Environment Act are the pillars of the Swedish regulatory framework on working life issues. Their practical ramifications are narrowed step-by-step, and can be specified down to *ad hoc* interpretations for local development projects. By way of example, the following excerpts from a management–labour agreement give a flavour of STA views on productivity and the introduction of new technology:

The STA is to attain the goals of its activities:
- with sound resource management,
- with due consideration for the requirements of service, publicity and legal security, and
- with due regard for the need of the personnel for job satisfaction, a good working environment, job security and opportunities of codetermination and personal development.

System design must be guided by the demands made on systems by their users, and the users are therefore to be given an opportunity of actively participating in the definition of requirements.

The aim of all computerization must be for systems to be designed in such a way as to create interesting and varied duties for the users of the system. Terminal work must be organized and work places designed in such a way as to comply with current directions and recommendations, i.e. so as to achieve a good working environment. (Codetermination, 1982)

It should be stressed that the above text actually represents joint management–labour views and is in no way mere lip service to 'the Swedish model'. It should also be stressed that there can be no sustained production without control and feedback at all levels of the production organization. In Sweden, decisions regarding

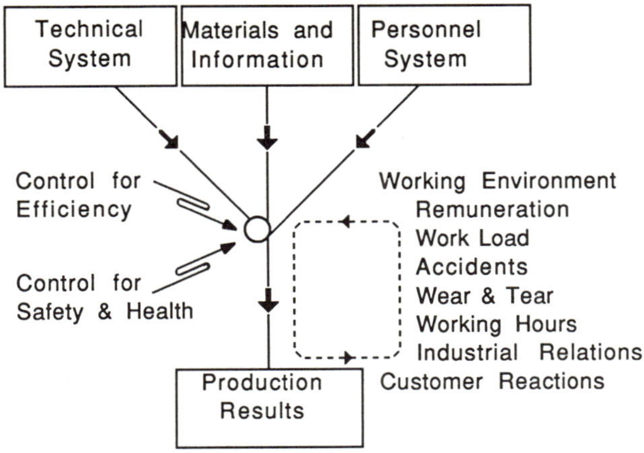

Figure 8.1 Like any system, a production system is not stable over time unless there are mechanisms for control and feedback. Decisions regarding these issues are concerns for management and labour alike. Note that the figure is not complete in that the system is open and only one internal feedback loop is indicated.

the breadth, depth and processes of such control and feedback are made jointly by management and employee representatives. Some 95% of the nearly 50 000 STA employees belong to one of three trade unions. These unions are acknowledged by the employer as representing the human bodies and minds comprising the 'personnel system'. The STA codetermination agreement and the parallel working environment agreement recognize the crucial role of personnel organizations in overseeing the control and feedback process outlined in Figure 8.1.

The case of crew size considerations

An omnipresent STA problem will serve to illustrate the control and feedback issues indicated in Figure 8.1. No organization can live on control and feedback alone; these are not final results but inputs toward the production proper. For the STA telephone directory services, the focus of the present treatise, the purpose is to attend to customer calls on 'What's the number for XXX?' and 'Whose number is YYY?'

There is a price tag for each such call. At present, the first service is not charged to individual callers but rather adds to the average call charge/monthly fee, whereas callers to the second service are charged approximately US$1.50; over 90% of the total STA costs for these services are for staff. The number of operators is calculated using a 70-year-old mathematical queue theory for telephone traffic (e.g. Erlang, 1917), which specifies the required number of operators for given average values of traffic intensity, operator occupancy, customer waiting time and customer service time. It is crucial, from a production efficiency perspective, at any point of time to have the directory services optimally staffed. Too high a staffing level is obviously costly. In the long run, too low a staffing level will also result in increased costs due to

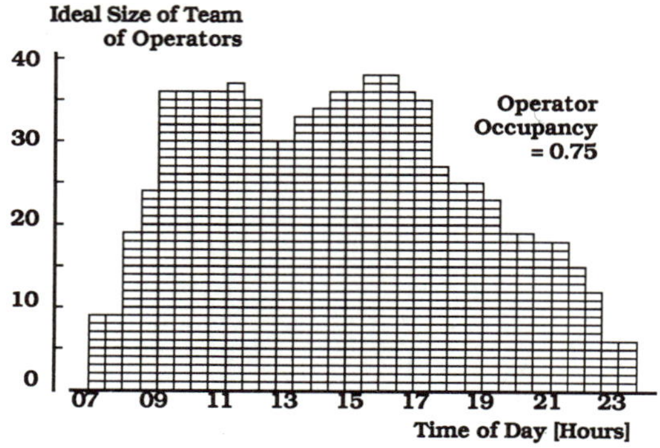

Figure 8.2 A typical time-of-day pattern of traffic intensity as related to the number of operators needed to ensure an operator occupancy of 0·75.

impaired quality of services (complaints from the customers) and quality of working life (complaints from the employees).

The chosen service level has a direct bearing on many of the factors in the control–feedback loop indicated in Figure 8.1. As part of production management, measurements are performed on traffic intensity, customer waiting time and time to serve a customer. In hour-to-hour operations management, these data must be compared to rules concerning working hours, rest-breaks, and workload. As seen from an outside perspective, the operators have to be monitored so as not to work too little (*efficiency control*) and also so as not to work too much (*safety and health control*). Important issues are, of course, precedence of one type of control over the other and the degree to which the operators themselves influence in the control–feedback loop.

Figure 8.2 makes the problem somewhat more concrete. It shows a typical time-of-day variation of incoming customer calls. The figure actually shows the traffic for STA directory service operations in a region of southern Sweden during a Monday in February 1988. The pattern is similar every weekday, but the level of traffic varies considerably between days, weeks and months.

Altogether there are 837 half-hour blocks in Figure 8.2. In theory, with operators working 8 h per day, this amounts to about 50 operators. In reality, since an operator cannot be split up in 16 independent half-hour units, about 90 operators are hired to do the job. Additional factors include time needed for scheduled and unscheduled rest-breaks, training, teaching, production meetings, vacations, overtime compensation and sick leave. There are also the traditional balancing problems of keeping all operators equally busy, which cannot fully be solved by either hiring part-time operators or switching operators between different types of jobs. The handling of all these factors has a direct bearing on the operators' working conditions. The most important determining factor, however, is the decision to design the operator team so that the probability of an operator being occupied is 0.75 (i.e. work being performed during 45 out of 60 min).

Ergonomics and productivity enhancement

If the design of the operator team had been based on a higher operator occupancy, then fewer operators would have been needed. Unfortunately, this increase in productivity would also mean that the average customer would have to wait longer to be served by an operator. Similarly, a productivity increase achieved by the operators trying to 'rush' the customers would be counteracted by a concomitant decrease in service quality.

The local goal of efficiency must not interfere with the global business goal. With reference to the feedback loop of Figure 8.1, STA runs the risk of getting caught in a *vicious circle* by stressing the above methods of local productivity enhancement. Fortunately there are other ways of increasing productivity, some of which at the same time increase service quality for the customers and working life quality for the opeators. Nobody can quarrel with such a *virtuous circle*.

Comfortable workstations

Figure 8.3 shows some STA workstation concepts over the years. It is readily seen that views on how to design and arrange workstations are intimately bound up with views on how to design, organize and supervise work tasks. The trend has been from supervised 'law and order' at standardized stations to self-governed work at ergonomically designed workstations.

The office with the round desk is from 1978. It represents STA's attempt to create optimal physical ergonomics as well as recognizing the importance of groups and teamwork. A lot of ergonomics had gone into the design of displays, desks, keyboards and lighting. However, there was a clash between the desk design criterion 'adjustable to operator size' and the job design criterion 'group work'. Birds of a feather had to sit together, because the big round desks had to be set for either short, middle, or tall operators.

The interior from 1988 is the one currently employed. The depicted desk costs around US $2000. It has been developed within STA and is available on the open market under the appropriate brand name ComforTable® (Östberg *et al.*, 1984). Among its features are pushbutton, electrically-operated height adjustment for the table top and a digital display showing the operational height in centimetres. This workstation meets the ergonomics criteria put forward by telephone directory assistance operations in, for example, the USA (Shute and Starr, 1984) and the Netherlands (Vellinga, 1987). It has become indispensable at the STA operator services operations, where the traffic situation shown in Figure 8.2. cannot be managed economically unless the pool of operators has access to a pool of easily adjustable workstations.

High-quality VDTs

Once requirements such as system compatibility and response time have been met, user-friendly dialogues and other *software ergonomics* issues can be addressed relatively independently of *hardware ergonomics* issues. Good hardware ergonomics, rather than

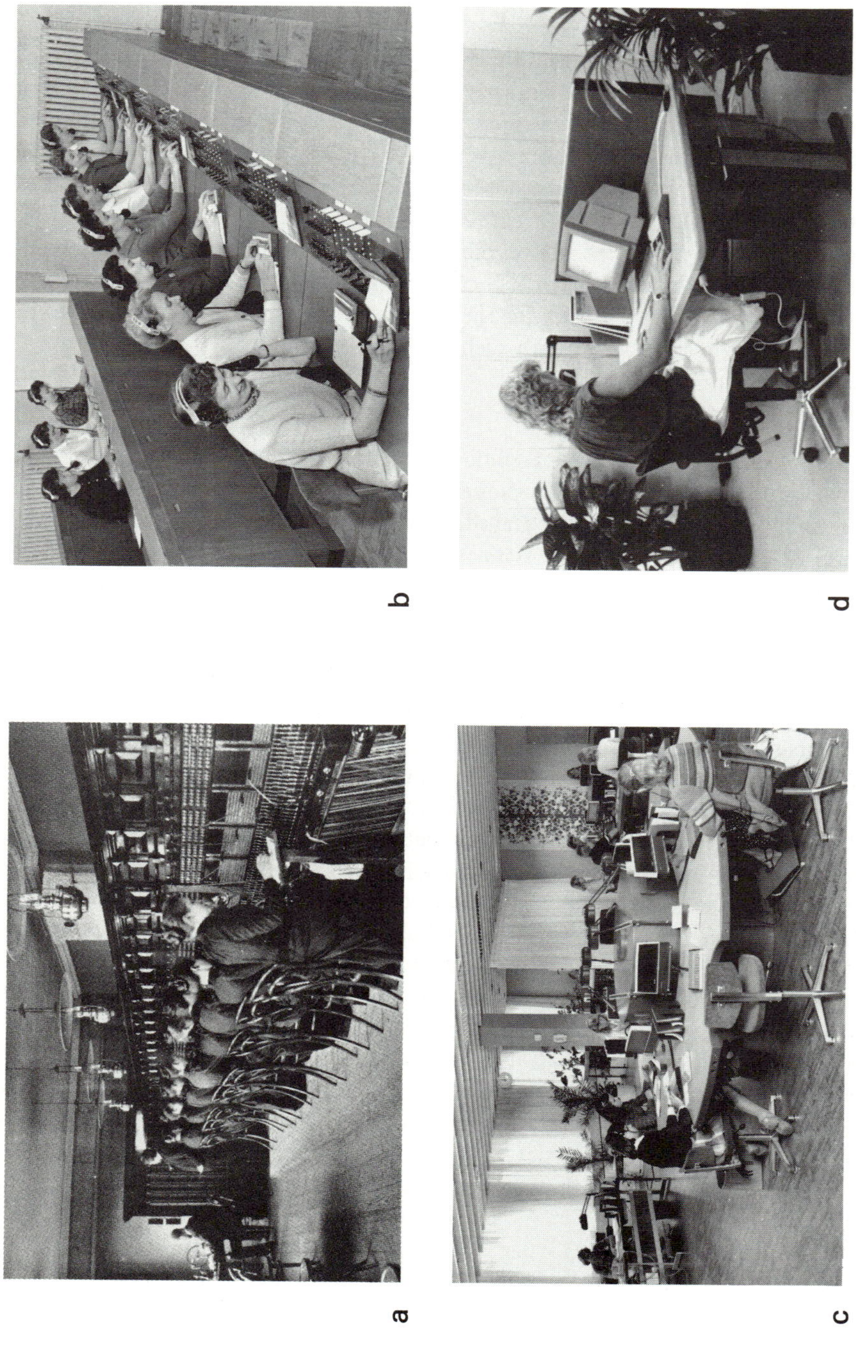

Figure 8.3 STA switchboard office interiors in the years (a) 1891, (b) 1957, (c) 1978 and (d) 1988.

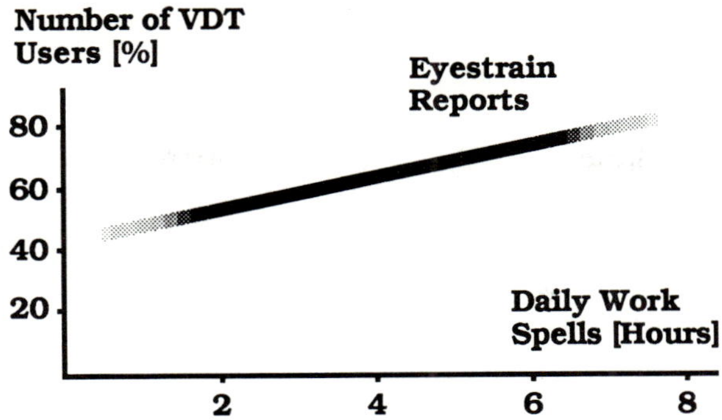

Figure 8.4 Complaints of eyestrain and/or vision problems. Data compiled from seven Swedish investigations of approximately 4000 VDT users reported in Östberg (1983) and Statshälsan (1988).

low cost, characterize the VDTs used in STA production offices. Thousands of new VDTs are now installed each year within STA, which gives the 'ergonomics' some leverage in the purchasing procedure (Östberg *et al.*, 1986).

Good operator–workstation interface is dependent on good desk–VDT interface. A design criterion for ComforTable® is that it should correspond to and match the VDTs it is used for. Adjustability not available in the desk has to be available in the VDT, and vice versa, so that the integrated workstation has good *physical ergonomics*.

With regard to *vision ergonomics*, such as freedom from flicker, eyestrain and visual fatigue, the VDTs have to be of high quality because very little adaptability and adjustability is available to the individual operator. The world over, operator complaints about the vision ergonomics status of their VDTs is commonplace. An indication of the Swedish situation is shown in Figure 8.4. Safety and health authorities over the years have come to acknowledge the existence of VDT related problems by issuing guidelines and minimum requirements. This is reflected in the VDT purchasing requirements such as 'black characters on a white background' and 'non-interlaced refresh rate of at least 70 Hz'. Furthermore, adopting MTFA ≥ 7 as the purchasing criterion for image quality would disqualify some 40% of the VDTs currently marketed in Sweden; according to STA not an unreasonable result (Feng *et al.*, 1990). By the way not a single US-made asynchronous terminal met the 1989 ergonomics requirements in effect for Swedish government agencies.

As a user organization with concerned line managers and keen safety and health departments, STA naturally addresses the issues related to problems indicated in Figure 8.4. As outlined in the aforementioned STA agreement, reduced operator discomfort is a goal if for no other reason than that discomfort sooner or later will result in performance decrement. However, a dose–response relationship similar to the one in Figure 8.4 can be found for any type of intense visual work (e.g. microscopy, on-line industrial quality inspection and intense hard copy reading), and exchanging *high-quality* VDTs for a paper file media will not in itself introduce any work-related eyestrain.

Support for this view comes from a recent IBM report. After many years of speed and error comparisons between VDT and paper reading, researchers at IBM finally concluded that the 15–25% performance decrements for VDTs could not be ascribed to any single vision ergonomics factor (Gould *et al.*, 1987). What was needed was a general quality shape-up to make characters displayed on a VDT as similar as possible to black characters printed on white paper. A multi-faceted approach (incorporating physical factors, work organization and visual factors) is being employed for VDT usage within the STA.

Microergonomics vs. macroergonomics

> Knowledge about the micro structure of steel alloys is necessary for the design and manufacture of light and strong steel wheels. That knowledge, however, is of little help in predicting the macro behavior of a steel wheel rolling down a hill. (Adapted from Sperry, 1969)

STA operators frequently alter their sitting postures by adjusting the height of their ComforTable® desktop. A new model is currently being introduced, in which the range of height adjustment is wide enough to allow the operators to switch to standing work. In general, the STA has gone to great lengths in using ergonomically designed workstations so as to reduce as far as possible the risk of wear and tear diseases from repetitive and one-sided work movements. These approaches, taken in isolation, might be categorized as *micro*ergonomics. Another micro approach would be to focus on measures for coping with a problem. An example would be to focus on the type of 'cure' in which the employer facilitates and encourages the operators to perform relaxation exercises during rest breaks.

In contrast, as outlined by Imada (1987), *macro*ergonomics deals with long-term ergonomics at an aggregate level. Both types are needed, however, in that micro measures are part of the macro strategy. This is in accordance with the ergonomics views manifested in the Swedish mandatory safety and health ordinances asserting that stereotyped work movements as well as strenuous and monotonous work must be limited in the first place. For example, *if* job design or workplace design measures fail to prevent strain from VDT work, *then* the employee shall be given adequate breaks in the course of work (NBOSH, 1985).

Ergonomics knowledge primarily consists of fragments at the micro level. The approach adopted by the STA is to integrate the various measures into a macro-ergonomics programme. The components of the STA rationale are as follows.[3]

1. Ordinances and recommendations from the Swedish National Board of Occupational Safety and Health speak in favour of well-balanced jobs.
2. It is the view of STA that 'the work should be organized in such a way that as many as possible share responsibility for ensuring that tasks classified as routine work are carried out' (Olsson and Thorstensson, 1986).

[3] The numbers are only used to enumerate the various components and do not indicate order of importance.

3. Adjustability of the workstation and the VDT equipment are necessary requirements for an operator to be able to switch between different work tasks.
4. High-quality VDTs with dark characters on a white background allow switching from VDT to non-VDT tasks without adjustment of the general office lighting level. Otherwise VDT work has to be performed at a lower level of lighting than that required for general office work.
5. Flexibility in the work organization and the work setting are necessary requirements when the traffic situation and the staffing level call for on-line balancing (compare Figure 8.2).

Flexible working hours are liked by the employees because they allow the individualized 'balancing' of job/family/leisure time. Such flexibility is the rule within STA, but until recently it was judged a problem for the customer services departments. This hindrance is now being removed.

Flexible working hours benefit the employer for industrial relations reasons. Furthermore, if the employer wants to benefit from employee initiative, creativity and problem-solving effort, then the employees must be entrusted with latitude for a variety of decisions, as well as opportunities and support for personal growth. The STA is a technically advanced organization and as such is undergoing a rapid and continuous evolution in order to stay ahead of the competition. It would be a very serious mistake to try to evolve without the involvement of the employees at all organizational levels.

Self-monitoring

With reference to Figure 8.2, traffic monitoring is needed to ensure satisfactory cost-effectiveness in speed and quality of the customer services. It is not necessary, however, to carry out electronic monitoring of individual performance on-line in real time so that minute-to-minute outputs can be tracked. In accordance with a STA management–labour agreement of 1981, the monitoring instead takes the form of aggregate systems measurements only. That is, no automatic performance monitoring of individual operators takes place. At each directory services unit, the operators collectively respond to the performance statistics displayed on blackboards. The performance is also continuously being discussed during regularly held production meetings. The operators' interest derives from the general need for feedback, the satisfaction arising from doing a good job, and that a productivity bonus is available within each production unit. The STA 'self-monitoring' system seems to agree with the following conclusions:

> It seems clear that electronic monitoring of employee performance is a necessary element in a competitive, productive work system. . . . Proper application of electronic monitoring can enhance job design by building intrinsic motivation into work activities. This is typically accomplished by increasing worker control over the task through worker participation in setting goals and establishing work standards and by providing feedback that assists the worker in gaining control over task activities, leading to satisfaction from successful performance. (Smith, 1988)

> Computer-aided monitoring is a phenomenon that is likely to become more prevalent in the workplace, and thus central to understanding contemporary employee response to work. This study develops a model for examining the impact of monitoring on employee-level job satisfaction and turnover. . . . The first implication of these results is that variation in turnover in automated offices, after controlling for traditional variables, is largely attributable to two quite distinct factors, (1) employees' affective responses to monitoring, and (2) prior beliefs or disposition toward monitoring. . . . Since managerial efforts are likely to have little or no effect on [the predisposed] employees the alternatives are to (1) screen out job applicants who are opposed to monitoring, (2) transfer to non-monitored jobs incumbent employees who are opposed, or (3) find alternatives to the use of monitoring. (Chalykoff and Kochan, 1989)

It should be pointed out that Swedish workers exercise decision power at multiple levels, and that management's only viable solution is to 'find alternatives to the use of monitoring' (Chalykoff & Kochan, 1989). This can be illustrated by the above STA example. First, at the off-line macro level, the operators had a say about whether or not monitoring should take place at all. Second, at the off-line micro level, the operators had a say in decisions about what type of monitoring should be implemented. Finally, at the on-line micro level, the operators have a say in how to use the monitoring information in resolving kinks in the production flow (e.g. who shall do what, where and when).

Visitors from foreign STA equivalents, particularly from the USA, find it hard to believe that this degree of job decision latitude works. The joint labour–management explanation is that there are virtually no serious decision latitude abusers among the employees, that the system thus works well, and that the job decision latitude ought to be deepened and broadened. The view of one of the STA trade unions is that an increased decision latitude would make possible a reduction from 8 to 6 h work per day. The argument is that the operators themselves know best how to utilize their personal resources so as to keep the production unit on target. Management is sceptical and considers this only to be an argument for 8 h pay for 6 h work, which, from their perspective, is a complete impossibility.

Connected to the union push for shorter working hours is the agrument that it would result in decreased exposure to the various electromagnetic and electrostastic fields emanating from VDTs equipped with cathode ray tubes. Although there is no scientific proof of any such VDT related hazards,[4] industrial relations considerations have led to the STA decision (i) that new VDTs for the directory services operations will have to meet quite stringent requirements on electromagnetic and electrostatic field strengths,[5] (ii) that existing VDTs will be rebuilt to meet these performance requirements,[6] and (iii) that upon a STA initiative the Stockholm regional services centre will be hosting a cause–effect study performed by researchers from the National Institute for Working Environment.

---

[4] But there is certainly a need for more research — see Goldhaber *et al.* (1988).

[5] Below 500 V surface potential, and below 25 $\mathrm{mT\,s^{-1}}$ at 0·5 m as measured by MPR (1987).

[6] At a cost of approximately US $500 per VDT.

A case study of VDT implementation

According to the aforementioned agreement on the operational use of computers, 'systems are not to be put into production until approved tests have been carried out in collaboration with users and data operations personnel'. Below follows an account of the reality behind such language. The example is from the second wave of computerization of directory services operations, initiated in 1980 at the Malmö regional centre.

The introduction of computers in support of old or new functions is a continuous phenomenon in the telecommunications business. The first generation of computerized information services was the round-table office at the Stockholm regional centre (Figure 8.3). As the resultant productivity was found to be significantly higher than in the manual operations based on traditional paper files, management wished to press on with further computerization. At the same time, an independent research team concluded that there might be some visual problems associated with extended VDT work at STA and that organizational factors such as work-spell duration and the rigidity of work routines seemed to play an important role in the frequency with which visual strain occurred (Gunnarsson and Söderberg, 1983).

The frequency of operator complaints within STA was some 10–15% lower than the general trend in Figure 8.4. None the less, operator comfort was an important consideration in the design of the system to be tried out in the Malmö regional centre. Great care was taken in the selection of VDTs and the design of workstations and working environment in the new directory services offices. During this process, as usual, union representatives and operators took an active part.

There were negative union sentiments toward the computerization. Specifically, reference was made to the operators' subjectively reported eyestrain. It was jointly determined that the objective eyestrain indicators employed by Östberg (1980) and Gunnarsson and Söderberg (1983) should be used as a part of the evaluation of the field trials. The investigation tools were (1) a questionnaire about visual strain, (2) a near point of accommodation rule, (3) a near point of convergence rule, (4) a laser optometer for assessments of dark focus and dynamic focus accuracy, and (5) a photometer for luminance and illumination measurements. A full account of this research has been reported elsewhere (Shahnavaz, 1982; Shahnavaz and Hedman, 1984; Hedman and Briem, 1984, 1987).

Ergonomics evaluation methods

Based on trade union demands, the rule '15 min rest break after 45 min VDT work' was strictly adhered to during the field trials. Another union demand was that 'intense VDT work shall not be performed for more than 2 h per day'. This second demand resulted in a two-phase study design, in which directory services work in the new VDT office (termed *VDT*) was interspersed with work in the old office for manual directory services (termed *MAN*) and mixed, non-VDT switchboard work in the office for cellular telephone services (termed *SWI*).

A total of 29 operators (27 female and 2 male) served as field trial subjects (mean age

27 years). Prior to the trials, a team of eye specialists tested their vision, and, when needed, prescribed corrective lenses.[7] The investigation period spanned three (Phase 1) plus three (Phase 2) months. All subjects had a basic 8-h *MAN* working day. They were scheduled to take part in measurements for eyestrain indicators during two (Phase 1) plus three (Phase 2) test days. The subjects served as their own controls.

During Phase 1, the test days consisted of a 2-h pre-lunch test session (*VDT* or *MAN*), a 2-h post-lunch test session (*MAN* or *VDT*) and a 2-h afternoon *MAN* session. The rest of the test day was devoted to tests, snacks and rests. During Phase 2, the test days consisted of a 3-h pre-lunch session, a 3-h post-lunch test session and 2-h devoted to tests, snacks and rests. The test sessions during the first and second test days were spent at either *MAN* or *VDT*, and the test sessions during the third test day was spent at *SWI*.

Field trial results

It was acknowledged by the researchers that the experimental conditions were consistent with the management–labour agreement (e.g. with regard to the generous rest break rule). The researchers initially concluded that the VDTs, workstations and environmental factors in the new VDT office met all general textbook ergonomics requirements.

The researchers' most important conclusion was that the results showed an almost total lack of differences in eyestrain indicators between *VDT*, *MAN* and *SWI*, both within and between the two experimental phases. Non-significant differences can always be explained by the use of an unsuitable test battery. The researchers, however, did find differences with regard to other factors.

For example, it was found that the precision of the subjects' accommodation was dependent on age and whether or not they were in a job rotation schedule (i.e. Phase 1 vs. Phase 2). All age groups showed over-accommodation when focusing at distant targets, and under-accommodation when focusing at close targets.[8] This was most pronounced for the older subjects and during Phase 1. Based on these results, the researchers concluded that a useful eyestrain index could indeed be defined with the test battery employed.

Conclusions

The *researchers* concluded that it was unlikely the work conditions studied would cause eyestrain. This was explained by the very high VDT and image quality, excellent facilities for workstation adjustments, job rotation and frequent rest breaks.

The *management* concluded that all concerned had done a good job in seeing to it that the STA computerization was characterized by high ergonomics standards, that no scientific or other reasons spoke against a nationwide extension of the Malmö trials, and that the computerization had proved to be profitable. Furthermore, customer service quality had improved significantly.

---

[7] This routine is free of charge for all VDT users within the STA.

[8] This problem and its possible remedies have been further investigated in a follow-up study (Östberg and Swanson, 1988).

Two of the three *trade unions* concurred with the management views and accepted full-scale computerization, given that the proportion of VDT work was kept to that of the field trials. The third union stressed that the subjective data of Figure 8.4 were sufficient grounds for restricting 'VDT intense' directory work to 2 h per day.

The point then and now (with several years of hindsight) is that the researchers were called in as part of the continual labour–management discussion on how best to promote productivity and health in the computerized office. The management pressed for a higher degree of computerization, and the unions agreed but pressed for 'profit sharing' in terms of shorter working hours, improved working environment and increased decision latitude. This process is continuous, and researchers can at best provide a common ground for the discussions. The view that 'planning for technological change should rely more on experts than end users' is listed as 'myth number 4' in the review by Östberg and Chapman (1988). On the contrary, for a rich account on the potential of participatory design approach, see Greenbaum and Kyng (1990).

Epilogue: on the politics of ergonomics policies

For all enterprises the local goal is efficiency and the global goal is efficiency/advantage/competitiveness.[9] If competitiveness cannot be achieved, companies will soon be out of business — and the STA is no exception. It is thus natural that the employer is the weighmaster in the continuing process of finding the optimum balance between the local and global goals. It is important to do things right (local efficiency), but it is also important to do the right things (global efficiency). It is important to use efficient computers (local efficiency), given that it results in, for example, better customer service at a lower cost (global efficiency), but it is also important to have efficient personnel utilization (local efficiency) and a motivated and skilled workforce (global efficiency).

The views of the employees are of course important in the evaluation of local/global issues, especially, as is the case at STA, if put forward in the form of responsible, collective views channelled through their own selected representatives. The ergonomics clout of organized labour is acutely revealed in Westin's (1987) field studies of US management policies toward ergonomics, health and safety, and work practices involving VDTs. In the period 1982–4, some 75% of the 110 progressive US corporations in the study showed 'weak attention' to ergonomics issues. Two years later, 50–60% of these did have 'serious and responsive' policies in place. At companies where the employees were organized, policy changes were the result of trade union pressure on management. At companies where the employees were not represented by a trade union, policy changes had been implemented so as to remove a reason for the employees to want to become organized.[10]

[9] Local/global refers to, e.g. task objectives, physical location and time perspective.

[10] In 1986 the overall percentage of unionized wage-earners was 86% and 18% for Sweden and the USA, respectively. The difference is considerably higher for office workers.

The proof of the pudding is in the eating. From a STA management perspective it does make sense to be 'serious and responsive' both with regard to the ingredients (microergonomics) and the baking process (macroergonomics). From a local perspective the approach painted in the present review may at times seem expensive and slow. From a global perspective, however, it is a very profitable approach and will help the STA to stay in business and remain profitable.

Similar views are held by a great many Swedish employers. In the book *The Doctrine Shift*, the think tank of the Swedish Employers' Confederation recently reviewed these issues and pointed to the necessity of even more far-reaching policy changes (Beckérus *et al.* 1988). The need for a doctrine shift is seen not to be the result of a wave of humanism or a changed power base, but rather a pragmatic codification of signals from the marketplace. The needs of employees and employers are converging. The employees demand both opportunities for personal and professional development and opportunities for expressing their views and exercising their judgement in meaningful work settings. The employers are in need of skilled, flexible and creative workers, who can act responsibly at all levels, take initiative and solve problems.

References

Beckérus, Å., Edström, A., Edlund, C., Ekvall, G., Forslin, J. and Rendahl, J. E., 1988, *The Doctrine Shift* (in Swedish). Stockholm: Svenska Dagbladet.

Chalykoff, J. and Kochan, T. A., 1989, *Computer-aided Monitoring: Its Influence on Employee Job Satisfaction and Turnover*. MIT Industrial Liaison Program Report, No. 4–44–89. (Cambridge, MA: Massachusetts Institute of Technology).

Codetermination Agreement for the Swedish Government Sector (MBA-S); Ratification comments by the negotiation parties of the Swedish Telecommunications Administration, February 18, 1982. (Stockholm: Swedish Telecommunications Administration).

Corlett, N., 1987, A tract for the times: what you could be doing within the next decade. *The Ergonomist*, (210), October, 1–2.

Deutsch, S., 1982, Unions and technological change — international perspectives. In: *Labor and Technology: Union Response to Changing Environments*, edited by D. Kennedy, C. Craypo and M. Lehman, pp. 191–201, (University Park, PA: Pennsylvania State University).

Erlang, A. K., 1917, Solution of some problems in the theory of probabilities of significance in automatic telephone exchanges. *Post Office Electrical Engineers' Journal*, **10**, 189–197.

Feng, Y., Östberg, O. and Lindström, B., 1990, MFTA as a measure for computer display screen image quality. *Displays* (in press).

Gardner-Bonneau, D. J., 1987, Process control: safety issues and answers in Sweden and Norway. *Proceedings of the Human Factors Society 31st Annual Meeting*, pp. 1243–1245 (Santa Monica, CA: Human Factors Society).

Goldhaber, M. K., Polen, M. R. and Hiatt, R. A., 1988, The risk of miscarriage and birth defects among women who use visual display terminals during pregnancy. *American Journal of Industrial Medicine*, **13**, 695–706.

Gould, J. D., Alfaro, L., Finn , R., Haupt, B. and Minuto, A., 1987, Reading from CRT displays can be as fast as reading from paper. *Human Factors*, **29**, 497–517.

Greenbaum, J. and Kyng, M. (eds.), 1990, *Design at Work*. (Baltimore, MD: Lawrence Erlbaum Associates.)

Gunnarsson, E. and Söderberg, L., 1983, Eye strain resulting from VDT work at the Swedish Telecommunications Administration. *Applied Ergonomics*, **14**, 61–69.

Hedman, L. R. and Briem, V., 1984, Short-term changes in eyestrain of VDU users as a function of age. *Human Factors*, **26**, 357–370.

Hedman, L. R. and Briem, V., 1987, Focusing variability during visual work. In: *Work With Display Units 86*, edited by B. Knave and P.-G. Widebäck, pp. 739–744. (Amsterdam: Elsevier Science).

Imada, A. S., 1987, Contemporary issues for macroergonomics. *Proceedings of the Human Factors Society 31st Annual Meeting*, pp. 173–175. (Santa Monica, CA: Human Factors Society).

MPR, 1987, *Testing Visual Display Units — Test Methods*. (Stockholm: MPR National Council for Metrology and Testing).

NBOSH, 1985, *Work with Display Units* (English issues of Ordinance AFS 1985:12). (Stockholm: National Swedish Board of Occupational Safety and Health).

Olsson, A. and Thornstensson, G., 1986, Organisation of work at data terminals. *Tele* (Focus on Swedish Telecommunications, English edition), **40**(1), 27–29.

Östberg, O., 1980, Accommodation and visual fatigue in display work. *Displays*, **2**(3), 81–85.

Östberg, O., 1983, Office computerization: research then and now. In: *Office Automation — Jekyll or Hyde*? edited by D. Marschall and J. Gregory, pp.127–142, (Cleveland, OH: Working Women Education Fund).

Östberg, O. and Chapman, L. C., 1988, Social aspects of computer use. In: *Handbook of Human–Computer Interaction*, edited by M. Helander, pp. 1033–1049. (Amsterdam: Elsevier Science).

Östberg, O. and Swanson, N., 1988, Visual accommodation when focusing targets in 'near' and 'far' areas of a chart containing a distance illusion pattern. *Proceedings of the 12th International Symposium on Human Factors in Telecommunications*, pp. F.4.1–8.

Östberg, O., Warell, B. and Nordell, L., 1984, ComforTable® — a generic desk for the automated office. *Behaviour and Information Technology*, **3**, 411–416.

Östberg, O., Möller, L. and Ahlström, G., 1986, Ergonomic procurement guidelines for visual display units as a tool for progressive change. *Behaviour and Information Technology*, **5**, 71–85.

Shahnavaz, H., 1982, Lighting conditions and workplace dimensions of VDU-operators. *Ergonomics*, **25**, 1165–1173.

Shahnavaz, H. and Hedman, L. R., 1984, Visual accommodation changes in VDU-operations related to environmental lighting and screen quality. *Ergonomics*, **27**, 1071–1082.

Shute, J. S. and Starr, S. J., 1984, Effects of adjustable furniture on VDU users. *Human Factors*, **26**, 157–170.

Smith, M. J., 1988, Electronic performance monitoring at the workplace: part of a new industrial revolution. *Bulletin of the Human Factors Society*, **31**(2), 1–3.

Sperry, R. W., 1969, A modified concept of consciousness. *Psychological Review*, **76**, 532–536.

Statshälsan, 1988, *Investigation of the Health and Working Environment of VDT Users* (in Swedish). (Stockholm: Statshälsan Foundation for Occupational Health and Safety for State Employees).

Törnqvist, B. and Hermansson, S., 1983, Automation and the quality of life at the Swedish Telephone Company: a management view. In: *Office Automation — Jekyll or Hyde*? edited by D. Marschall and J. Gregory, pp. 84–88. (Cleveland, OH: Working Women Education Fund).

Vellinga, R. 1987, A user-designed terminal table system. *Behaviour and Information Technology*, **6**, 353–361.

Westin, A. F., 1987, Trends in U.S. user policies for VDT work. In: *Work with Display Units 86*, edited by B. Knave and P.-G. Widebäck, pp. 670–681. (Amsterdam: Elsevier Science).

Westman, B., 1983, Automation and the quality of life at the Swedish Telephone Company: a union view. In: *Office Automation — Jekyll or Hyde*?, edited by D. Marschall and J. Gregory, pp. 84–88. (Cleveland, OH: Working Women Education Fund).

Chapter 9

Organizational Culture and VDT Policies: a Case Study of the Federal Express Corporation

ALAN F. WESTIN

Introduction

Good ergonomics in terminals, furniture and physical work environments and extensive employee health and safety education represent necessary, but not sufficient, policies to achieve safe, healthy, productive and employee-satisfying use of visual display terminals (VDTs). Without a comprehensive 'people-technology' philosophy, process and implementing environment, good ergonomics and good education will not only not be enough but may also delude many employers into thinking that they are addressing the use of office systems technology in responsible and effective ways.

The case study that follows describes how one company — Federal Express — went about developing excellent VDT policies in the 1980s, and how it continues to manage 'people–technology relations' adroitly today. The synergism of tool–worker–production–organization that Federal Express has achieved represents both a model of good employer policies and a proof that this can be done, with some difficult learning experiences along the way, in the very real-world, competitive environments that employers face in the 1990s. (A note on Sources for this case study appears after the text.)

Federal Express: the organization and its business

Federal Express, which began operations in 1973, is now a large multi-national company with fiscal year 1989 annual revenues of US $5·2 billion, 56 841 permanent employees and a well-known advertising slogan that invites shippers to use Federal 'when it absolutely, positively has to get there on time'. As the company described itself in its 1988 *Annual Report*, Federal Express 'specializes in the door-to-door express delivery of goods and information throughout the United States and many foreign countries,' using an 'integrated air–ground transportation system.'

In its 16-year existence, Federal Express has grown into a large, complex, world-wide operation. It has over 29 000 ground vehicles (vans and trucks), 300 aircraft, 1300 Federal Express offices in the USA and abroad, plus 17 000 'drop boxes', 400 business service centres, and 275 drive-through kiosks. *Each day*, in 1988, the company handled an average of 1·1 million packages and documents, fielded over 244 000 telephone calls, made over 350 000 courier delivery stops, and passed millions of items of information through its extensive computer network. In fiscal year 1989, the company collected and delivered a total of 270·3 million documents and packages, at an average on-time delivery rate of about 98·8%.

The Federal Express workforce divides into 8–10 major occupational groups: couriers driving its vans, dispatchers and station agents, pilots and ground crews staffing its 'air force', service agents handling customer contacts at its call centres, sorters at its regional Hubs, a professional-technical staff of salespersons, accountants, engineers, personnel specialists, lawyers, information technologists and others, and a cadre of line, regional and national managers and executives. Federal is a non-union company in all of its operations.

'Technological innovation is our lifeblood'

From its earliest days, Federal Express has applied advanced technology to the design, operation and constant improvement of its products and services. Into an industry — air freight delivery — that was highly traditional and backward in its methods in the 1970s, Federal Express brought its famous 'hub and spokes' system of package pick-up and air delivery each night to Memphis and then turnaround air carriage back to the points of destined courier delivery the next morning.

Information technology — computers and telecommunications — has been and remains basic to Federal's system. Over the past decade, Federal's innovations in information technology have included:

1. COSMOS — A computerized tracking system that monitors the flow of packages through the Federal Express system and locates them for customer inquiry.
2. DADS (Digitally Assisted Dispatch System) — Mobile data terminals in the courier vans (also used as portable units by on-foot couriers), these data terminals transmit package pick-up and delivery information and greatly reduce the need for telephone communication.
3 SUPERTRACKER — A portable mini-scanner device that sends package information through the DADS terminal directly into the main computer network for real-time data transmission.

Though it has been committed to 'high tech' information and communication systems for its competitive edge, Federal Express management has insisted that its technological innovations satisfy three 'people-first' requirements.

1. Innovations must directly improve customer service and satisfaction, not be 'technology for its own sake'.
2. Technology must bring out the best and most creative aspects of its employees, not try to substitute machines for things workers do best.
3. Use of technology should not undermine Federal Express's promise of 'job security' to its employees.

In developing its technological innovations throughout the 1970s and 1980s, these concepts have been primary forces in Federal's system designs and applications. After describing its latest technological innovations in its 1988 *Annual Report*, Federal Express told shareholders and potential investors that it saw the *effectiveness* of its technological programmes resting on its employees' readiness to embrace these new tools. 'The best technology in the world is useless, of course, unless it is in the hands of a well-trained and highly motivated workforce. Maintaining high levels of job satisfaction has always been a paramount concern at Federal Express, a company whose personnel policies have generally been considered among the most progressive and innovative in America.'

In short, when office systems technology and the VDT arrived in the late 1970s and early 1980s, Federal Express had in place a comprehensive human resources approach and a clear 'people–technology fit' concept that allowed it to recognize, explore and deal effectively with the new issues of using VDTs in its operations.

VDT uses at Federal Express in the early 1980s

Federal Express began using word processors and microcomputers in the late 1970s. By 1983, Federal Express had over 8000 VDTs in use. About 1200 were in Federal's three large customer service telecommunication centres, then in Memphis, Tennessee, Somerset, New Jersey and Sacramento, California. About 250 VDTs each were used in the company's Treasury and Credit Department and its Revenue Accounting unit, primarily for data-entry work. Almost 120 VDTs were used in the Information Central unit. The remaining VDTs were spread throughout the country in station operations and in professional-executive use.

About mid 1983, Federal's Director of Human Resources Analysis, Dr Charles Hartness, became aware of the unfolding public controversy about VDTs — especially health and safety considerations and quality-of-worklife impacts from intensive VDT jobs. Hartness concluded that 'people–technology' issues in the application of new office systems technology could also affect Federal's employee and labour relations environment in a highly negative way, unless the firm managed these uncertain issues carefully and well. The problem, Hartness felt, was to develop a pro-active approach to the VDT issue, based on three perceptions: (1) this is *not* a narrow 'ergonomics' issue involving equipment and furniture features; it requires rethinking job design, work-group relations, work quotas and performance supervision, employee education and participation and *manager* education; (2) Federal Express should identify what other pro-active peer companies are doing, and draw on expert consultants who work with 'leading edge' firms; and (3) the policies developed for VDTs must be formulated and institutionalized with strong support from Federal's founder and CEO, Fred Smith, and linked with the future of both technological innovation and employee job security. Hartness sold these ideas to Smith, and received a mandate to develop a 'comprehensive VDT policy'.

After considering ways to organize this response, Hartness decided to create a special VDT task force, led by the Human Resources Department, to consider what actions would be necessary to insure that the Federal's 'people-first' philosophy governed VDT applications.

Forming the task force, setting the agenda

Federal's VDT task force held its first meeting in July of 1983; it met regularly until late 1987, and has met occasionally in 1988–9. Depending on the agenda, anywhere from 8 to 20 participants attended the meetings in the 1983–7 period, and minutes of the task force activities and decisions went to over 50 staff and executives.

In its first half year, to the end of 1983, three groups within the company provided its membership: *human resources* (experts from corporate safety and health, human resources analysis, group insurance, organizational development and training, and personnel); *operations with extensive VDT use* (staff members from customer service, treasury and credit, and information central); and *communications* (staff members from the employee communications and corporate communications groups). Later, representatives were added from more user departments, and also from information systems, properties and facilities, management communications, dispatch management, corporate finance and many other functions. By 1986, nine user departments and 23 staff groups had participated in the VDT task force work, giving the operation wide 'reach' and impact within the company.

For chairman of the task force, management selected Fred Rine, Corporate Director of Safety and Health, a unit within the Human Resources Department. This appointment cast the VDT issue initially as a 'scientific inquiry' to fulfil corporate health and safety responsibilities toward employees. This focus gave the mission a high-priority and high-status 'launch' within the company, rather than an 'equipment and furniture' focus.

By the second meeting in September 1983, the task force had drawn up a 'video display terminal safety and health programme' with no less than 16 items specified for action. These included:

(a) immediate upgrading of ergonomic conditions at some terminals and workstations,
(b) reconsidering job design for intensive VDT tasks to avoid monotony and reduce stress,
(c) redesigning physical environments for VDT operations,
(d) formulating new performance evaluation criteria for data entry and customer service work,
(e) instituting a stress-reduction oriented exercise programme for VDT operators and a rest break policy for intensive VDT tasks,
(f) re-examining compensation and benefits programmes to insure proper pay and incentives,
(g) addressing concerns about the safety of pregnant VDT workers and formulating an employee communication programme on VDT issues,
(h) developing a major new employee vision and hearing care programme, as a health-care benefit,

(i) addressing ergonomic features with Federal's prospective workstation and work-facilities vendors, and
(j) beginning work on a company-wide ergonomic standard to submit to top management for funding and implementation.

From our current perspective in 1990, we can see that this was a remarkably comprehensive agenda of issues for 1983, precisely the comprehensive human resources oriented approach that is critical for managing the physical and mental aspects of stressful high-tech office work. It was also right on the money, in terms of effective VDT policies that other pro-active user organizations would turn to several years later, in 1985–7, but *which a majority have yet to embrace in 1990*!

While some important additional actions were added as the task force proceeded, the remarkable aspect of Federal Express's VDT approach is how well its task force conducted research, assessed alternatives, framed its basic 1983 agenda, promulgated it throughout the company and has kept it updated ever since.

We can examine this approach by looking at four main areas of task force action:

1. gathering of information about VDT issues and employee concerns;
2. treatment of job design and performance evaluation issues;
3. development of VDT ergonomic standards; and
4. manager and employee communications programmes to foster sound VDT *practices*.

Assessing VDT conditions and employee concerns

The task force began by sending members out to visit other major users of VDTs — an insurance company, airline and credit-card company — and also a major terminal vendor. Scientific literature was collected on radiation, musculoskeletal issues, stress, vision and other health and safety issues, and leading consultants in these areas were brought in to confer with the task force. Meanwhile, a team visited each of the company's six main VDT-using operations, examining terminals, workstations and work environments, and noting ergonomic shortcomings, especially in the older facilities.

To see how Federal's VDT operators were experiencing the technology, and how publicity in the mass media might be affecting their attitudes, the Communication function conducted 'Focus' sessions with small groups of employees at seven different sites using VDTs; the discussions were quite frank and free-wheeling. While concerns varied by the type of employee group and VDT task, a set of broad general concerns came through in the sessions. These involved eyestrain, radiation worries, ventilation, equipment maintenance, poor chairs, rest breaks and better information and training. A report describing and analysing these employee concerns was presented to the task force, which in turn sent a memorandum on the main items to Federal's top management. A similar focus group exercise was done in 1984, as part of planning for a new data-entry system known as POLARIS.

In 1985, finding the information from the focus groups so valuable in policy considerations, the task force decided that direct employee participation in its proceedings would be both useful and practical. From that point on, employee representatives from the Memphis call centre, POLARIS group, Information Central, and the dispatch management attended task force meetings.

The task force also retained a small consulting firm, Changing Workplaces (of which the author is President), to provide input as to what other innovative companies and government agencies were doing about VDTs; to supply continuing materials on health and safety issues, advocacy-group positions, legislative and legal developments, and available national employee-survey data on VDT experiences; and to put on a series of seminars and tutorials in Memphis between 1983 and 1987, first for managers and professional staff and then as annual events for interested managers and employees (these are described later).

Reshaping jobs and evaluating VDT work performance

As the task force sorted out the real issues in VDT work and discarded those not supported by solid scientific evidence or empirical studies, the issues of good job design and performance evaluation emerged as significant matters. This was especially true for VDT work in the 'pressure cooker' environment of Federal's daily operational deadlines.

First, Dr John Hater, an industrial psychologist in human resources, studied the way that several hundred data-entry jobs had been planned in the new POLARIS operation. He concluded that these would be too monotonous and stressful if they were conducted as straight VDT work throughout the day. On his recommendation, the POLARIS team combined some non-VDT work with the VDT tasks, with the expectation that optical scanning and other methods would probably phase out the POLARIS jobs by the end of the 1980s. (With the installation of hand-held scanners as input devices in the courier vans, full-time data-entry units have largely disappeared at Federal Express.)

Secondly, the task force became centrally involved in a collision between operational units and human resources executives over how to monitor and measure the performance of customer service operators at the call centres. The company in 1984–5 was facing stiffened competition in the overnight delivery business, and the impending entry into this field by United Parcel, long regarded by Federal's top management as its most serious future challenger. Costs had to be cut and efficiency strengthened, and cost-controls dominated much of call centre administration during this period. In 1985, the senior manager in charge of the customer service function set out to improve performance by achieving a reduced 'average handle time' for customer calls. To do this, he told supervisors to rely on the customary occasional system of monitoring of employee calls and the assignment of a 50% rating of performance on this 'quantity' dimension.

Human resources executives in Memphis soon became aware of rising employee discontent over this new enforcement policy. A consultant (Westin) was asked to conduct focus group sessions with customer service agents and line managers at the

Memphis centre. The following excerpt from the consultant's report described what was found to be happening:

Major Themes Voiced by the Agents

1. There was a high anger level over the current work standard being 'imposed' on them — the 140-s average handle time, and the enforcement mechanisms (supervisor monitoring and performance appraisals) that were being applied to make the 140 s standard stick. They were especially angry that experienced and excellent agents were being forced to take 'retraining' if they exceeded the new average, and that agents that did not have even lower average handle times would not receive the high performance ratings that opened up transfers to better jobs and promotional opportunities.
2. Agents feel they can no longer give quality service to customers under this standard. Extensive examples were supplied of having to cut callers off, pass them on to places where their questions would not be well handled, deliberately avoiding solving problems they once would have undertaken, not promoting additional services, etc. 'Making the 140 seconds' is now 'the dominant thing in our minds from plug-in to plug-out'.
3. Agents are convinced that FEC will lose business to competitors and alienate long-standing clients because of the lowered quality of service they are now forced to supply. Examples were given of customers already complaining to them about 'the old FEC courtesy and helpfulness' not being present any more.
4. They were 'astonished' that the new IBM terminals were introduced without a minute of training or any advance briefing, even though there are 'substantial differences' in keyboard and operation.
5. They were angry that no learning or adjustment time was provided for 'getting up to speed on the new terminals' (which took them 2–3 days), and that they were expected to make their 140 s from 'the minute that we got to work on the new machine.'
6. The chairs they have cause back pain and muscle tension, with one person having developed severe neck cramps and having to start a series of medical treatments to be able to stay at work. They need adjustable backs with proper support, arm rests, and several smaller women pointed out that they needed foot rests. The two groups said they had been 'promised new chairs 2 years ago' but 'nothing happened'. 'Now, you have to bring in a doctor's note saying you are in medical trouble before you get a proper chair.'
7. A major rise in stress-related conditions was reported by most of the agents. They described having stomach pains, sleeping problems and increased irritability and anger in their off-job lives. Several said they are now under their doctor's care for stress.
8. They voiced concern over their pay being inadequate for (a) the knowledge they bring to their work about the entire operation of FEC, and (b) the constant tension and 'health impairment' they suffer under the stress of their jobs (angry customers, time pressure, and 'the 140 s handle'). They also feel there is now little career opportunity for them — 'seniority means nothing,' and 'management wants to shrink our job rather than enlarge it.'

9. They said they are at a low point of morale as FEC employees; over the past 2–3 years, 'our jobs have gotten worse and worse.'
10. They are deeply upset at FEC management. 'They have taken the 'People' out of the system. Now it's the Numbers, and someone's idea of short-term gain over long-term service.'
11. The agents made it very clear that their concern was 'not with the technology' (despite some ergonomic problems). It is 'the way FEC management is abusing the technology' and 'driving people'. Sample comments were: 'They are treating us like children to be watched and punished.' 'It's just Big Brother monitoring all the time.' 'This company is selling out its most precious asset — the way we treat our customers.'
12. The agents noted that getting a job with FEC was something thousands of people in Memphis wanted, and that they were afraid to 'rock the boat too hard.' 'We have to work — we want to work at FEC — but management is making our jobs terrible. Satisfying customers is what makes the pressure of our job bearable, and management has taken that away from us.'

These reactions and feelings reported to Federal's management signalled a breakdown of Federal Express's people–technology approach at the service centres. What had happened was that key ingredients of Federal Express' stress-mitigating organizational climate had been allowed by this unit's management to fall away over a 2-year period, in pursuit of improved bottom-line performance. Employee involvement in standard-setting, good ergonomics, good training in a new terminal system just installed, employee satisfaction through superior client service, 'fair pay' for the pressure-work being done — all of these factors that had allowed VDT applications to proceed with strong employee support in the past — were turned from positive to negative through the catalytic action of the new 'get the handle time down!' enforcement campaign.

When these employee concerns were reported to CEO Fred Smith, correction of the situation was made a corporate priority. A new performance evaluation system was developed that put greater emphasis on nine 'quality' elements and used only one 'quantity' measure, in a change that was worked out with major input from a group of call centre employees. Financial authorization was obtained to buy top quality ergonomic chairs, and an employee test group made the decisive recommendation between the two final chair candidates selected by the call centre staff and the facilities group. (See Appendix for an article in the employee newspaper reporting this process.) Almost US $500 000 was spent in 1986 on these new ergonomic chairs, with an even larger amount scheduled to be spent in a second round of chair replacements at call centres in 1987.

In addition, training was provided on the new IBM terminals; an expanded exercise and adaptation programme for VDT work was instituted at the call centres; and an upgraded employee-compensation programme was developed. (The senior manager of operations was transferred to another function.)

The individual-evaluation monitoring of VDT work by supervisors continued at the Memphis call centre (and is still used at all Federal Express call centres today). This receives overwhelming approval from the agents. An anonymous survey of 587 employees done by Westin in 1987 found that over 90% felt that the supervisory monitoring programme then in force was 'about right'; should even be used more often; or was only 'a little too much'. Only 9·8% of respondents said that management was doing 'much too much' monitoring.

With Federal Express's people–technology principles restored, the entire situation at the call centres turned around in 1985–6. Employee morale was restored, and — as though to prove that virtue sometimes triumphs — the average handle time at the Memphis call centre was brought down by the new programme of employee–management co-operation *below* the level that had first been attempted by a coercive programme.

Similar employee-sensitive ergonomic steps were taken in other sectors where VDT use was expanding at the company. For example, employee surveys and focus groups on the VDT environment in dispatch offices led to testing mesh and glass glare guards for terminals, and selection of a glass guard for use in every office.

Developing a corporate ergonomic standard

Like most large VDT users, Federal Express found itself in 1983–4 with a mixture of different vendor terminals, desks, chairs, lighting arrangements and work environments. Conditions varied widely, depending on the type of operation, its age, whether the space was owned or rented and similar variables.

Once the task force had established that upgrading ergonomic conditions was a priority at many sites for both operator comfort and full productivity, it began drafting a 'corporate standard for VDT work.' This required analysis of key ergonomic elements, selection of criteria, and consideration of how such a standard could be financed and implemented by different departments.

Work on the standard went forward in 1983, and a recommendation was presented to top management during February 1984. This was approved and was officially promulgated as company policy in March 1984. The central goal of the 'recommended standards' was stated as follows:

> To insure the health, well-being, and commitment of those employees whose work environment and function are tied to a sedentary existence and Video Display Terminal (VDT) use.

The standard presented a discussion and 'guidelines' for seven areas: (1) equipment and workstation design, (2) image quality and display design, (3) lighting and reflections, (4) general workstation environment, (5) job design and organizational factors, (6) conclusions about radiation hazards, and (7) eye care.

Since the task force had already been helping user departments to implement many of the concepts in the recommended standards, issuance of the document in 1984 served primarily to give official policy status (and budget support) to carrying out the task force's approach to VDT jobs. In November 1987, standards for video display terminal use were incorporated into the *Federal Express Corporate Interior Guidelines Manual* (see Appendix).

Manager and employee communications on VDT issues

Between 1983 and the present, as the work of the task force progressed, a widening series of communications with managers and VDT end users was made a central part of the task force programme. A seminar on the sources and nature of the VDT controversy and how other leading organizations were setting new VDT policies was presented to about 60 key staff and management officials by an outside consulting firm (Changing Workplaces) in late 1983. A booklet explaining VDT health and comfort issues, prepared by the National Safety Council, was distributed to VDT operators in 1984.

In 1984 and 1985 articles on VDT issues and how Federal Express was approaching them were placed in the monthly *Manager's Pak* that serves as the company's primary written briefing for its entire management force. An April 1984 article, 'How Are Video Display Terminals Affecting Our Work Groups' provided a thorough briefing on VDT issues and related how the task force was approaching these within Federal Express. This article explained the unique nature of intensive VDT work, the harmful stress that can be generated, and the combination of measures that needed to be taken to provide sound VDT work environments. In noting the bad experiences that some companies had encountered, the article emphasized that such companies 'failed to communicate openly and honestly with their employees about VDTs, even after the employees raised the issue. They also failed to recognize and address discomfort concerns their employees had. Clearly, the existence of our VDT task force and our ongoing effort to address the issue in our company publications indicate we don't intend to fall into the same trap.'

In October 1985 an all-day seminar on managing the VDT environment was held in Memphis, with leading outside experts in radiation, ergonomics, performance and productivity, and employee relations aspects of VDTs. A mixed audience of 250 managers and employees attended the seminar, and a full report on the seminar's themes and conclusions appeared in the Federal Express employee newspaper. Additional seminars were held in 1986 and 1987.

Since 1986, VDT manuals have been distributed to employees, videotapes on effective use of VDTs were prepared and shown to employees, a 'stretch and flex' exercise programme was instituted, and the company-paid vision programme recommended by the VDT task force was provided (for all employees).

Pulling it all together: Federal's 1986 technology–people impact policy and its dissemination

What might be called the culmination of the VDT task force's first 3 years of effort came in 1986 with two highly symbolic actions:

1. The task force drafted and top management issued (in May 1986) an official 'technology–people impact policy'. This commits Federal Express

   > to systematically incorporate human factors, ergonomics, and job/task design criteria with the development or modification of electronic technology applications.

The policy specifies basic guidelines for workstations, hardware and software, job design, and implementation, and provides an impressive set of criteria for employee satisfaction and productivity in using VDTs. (See copy of this policy at end of this chapter.)
2. CEO Fred Smith and then Executive Vice President and Chief Operating Officer James Barksdale began, in 1985, describing Federal Express' VDT task force work and the 'people first' approach to technological innovation in major speeches that the two executives were presenting to business and public affairs groups. These speeches cited Federal's 'VDT policies' as a prime example of the company's organizational culture and management style, and a key explanation for Federal's achievements in the 1980s. By 1987, Federal Express was identified nationally as a 'leader' in the incorporation of VDTs into an efficient, 'employee-centred' work environment.

Stress-control and wellness programmes at Federal Express, 1986–9

In 1986, about the same time that Federal Express had brought its VDT task force programme into force, the human resources function was also beginning the design of a new company approach to its self-insured health benefits programme. Conceived as a comprehensive wellness-oriented approach, this would cover all employees, from customer service operators and couriers to professional staff and managers. It would also be extended to employee dependents, since Federal Express realized that family, home and community problems affected employee health and performance, and since dependents use of employee health benefits represented a major component of employer cost. This meant that Federal Express would be designing a new wellness programme to be used by 125 000 persons nationwide, not just its 50 000 employees.

The decision to create a new approach grew out of management's belief, as Corporate Safety and Health Director Fred Rine put it, that American business needs to 'seize control of sky-rocketing health-benefit costs,' rather than leave the primary direction of employee health utilization to the 'fee-for-treatment' based health-care establishment; '97% of the billions our society spends on health care goes for treatment and only 3% for prevention,' Rine observed. 'That's just wrong, and it is no longer a viable policy for companies like ours.'

Federal Express's new programme would emphasize prevention rather than treatment, and pro-active wellness policies rather than primarily remedial responses. Planning for the new wellness programme proceeded along the following assumptions:

(a) Experts at Federal Express were aware that workers compensation claims for stress were rising dramatically nationwide, employee utilization of health benefits for stress-related effects was increasing (25% of medical costs at Federal Express were for psychiatric treatment), and stress-related total-disability cases were also growing. 'We realized that if we didn't start to manage stress better,' Fred Rine recalled, 'it would manage us.'

(b) At the same time, given the company's inescapable daily 'pressure cooker' environment, there was no way that management could eliminate the stressful time deadlines, exacting documentation procedures, and challenging physical environment involved in getting deliveries there as promised. Rine observed, 'We defined our task as finding ways to help our employees use stress as a positive, direct their frustrations into healthy rather than unhealthy channels, and not allow stress to lead to negative attitudes and bad health practices.'

(c) New technology, including intensive VDT work, was not creating nor was it worsening stress problems at Federal Express. Rine commented, 'VDT use could have exacerbated stress and would have if we hadn't followed our 'people–technology' policy, adopted our VDT task force programme, and embedded this in our larger human resources approach. We know our problems are not based on the tools people use.'

(d) However, changes in using tools — especially advanced-technological tools — are central to progress and competitiveness at Federal Express. 'If people see change as harmful stress,' Rine noted, 'the company can't innovate rapidly. So we have to create an environment in which adapting to change, physically and mentally, whether the new procedures are technological or otherwise, is seen as a positive challenge.'

(e) Management was aware that it had a subset of its employee population at special risk and generating heavy health-care costs. About 15% of Federal Express's employees were using 85% of the health care dollar, and total utilization by employees was almost twice the national average of medical bills per employee for a comparable age-level workforce (30–31 years). What this called for, management concluded, was a programme to identify these extra-risk employees, not to punish them but to work with them in health education, prevention and rehabilitation. This would mean both better health for them and fewer dollars spent by Federal Express, along with better productivity.

The comprehensive wellness programme that Federal Express instituted in mid 1987 is called 'Spring'. So far, it has four main components.

1. *Education* — An extensive internal communication programme on identifying and overcoming health hazards, carried out through articles in employee publications, manuals, signs and educational meetings. Topics include smoking, obesity, stress, AIDS, high blood pressure, back problems and VDT use.
2. *Wellness* — Sponsorship of health-related intervention programmes, both short- and long-term. These range from providing exercise facilities on company property (gyms, fitness trails, aerobics classes, martial arts, etc.) to smoke cessation classes, weight watchers and Health Fairs on medical, hospital and dental issues. In addition, the company subsidizes membership in designated health clubs, such as 'Ys'.
3. *People help* — A nationwide employee and dependents counselling programme delivered by an independent private counselling firm, in a confidential setting. This help covers 'marital conflict, stress, relationships, alcohol, money worries, substance abuse, family problems and divorce'. About 80% of the cost of the Spring programme is centred in this employee and dependent assistance.

4. *Med-quest* and *options and choices* — A nurse-staffed telephone hotline and a medical consultant service to provide advice to employees and dependents on any medical problem. Depending on the nature of the problem, either the nurses or the medical experts describe risks and options involved in procedures recommended by the employee's or dependent's doctors, so that the employee or dependent can make better informed decisions.

The human resources function sold the Spring programme to management as a substantial project but one that would 'cost-justify' itself. An in-depth data collection programme was attached to Spring to insure that both short- and long-term evaluation of costs and benefits would be available. The programme was also sold to employees by explaining that every dollar spent unnecessarily on health benefits was a dollar not available for the employee profit-sharing programme. Thus, widespread and successful health-hazard prevention and reduced company health care costs directly enhance employee rewards.

The relation between stress and perceptions of fair treatment

The close relation between stress (in VDT work and elsewhere) and employee perceptions of fair treatment was demonstrated by an anonymous survey of Federal Express employees and managers conducted by Westin in late 1987. As part of a larger survey of employee relations issues and the operations of Federal Express's internal complaint-resolution system (called the guaranteed fair treatment procedure or GFT), 578 employee respondents and 483 manager respondents answered this question: 'If you felt that management was not treating you fairly on important issues, and if this was not remedied by discussions or the GFT, how much do you feel this would affect the following . . . ' The list of possible effects provided to respondents included (in this order):

Your productivity
Your attendance and tardiness
Your willingness to offer suggestions
The quality of work you would do
Your family life at home
Your stress level and health
Your remaining at Federal Express

Combining the responses of those who felt unfair treatment would affect an area either 'a lot' or 'somewhat', Table 9.1 shows that 'stress level and health' was chosen by managers as the most likely effect on them, and by employees as a close second effect.

The fact that more than four out of five employees and managers agreed that unremedied unfair treatment would increase their stress to levels that would affect their health is a powerful confirmation — directly from workers and managers — of the link between organizational climate and health effects; it is an insight that has

Table 9.1. Likely effects of unfair treatment at work: employee and manger responses.

Area affected by unfair treatment	Percentage response	
	Employees	Managers
Stress level and health	85·2	89·6
Willingness to offer suggestions	86·4	82·7
Productivity	79·8	80·2
Remaining at Federal Express	75·4	67·2
Family life at home	67·1	68·6
Quality of work done	59·5	54·5
Attendance and tardiness	38·1	27·8

animated Federal Express to see its employee relations philosophy, policies and programmes as major ingredients of its overall wellness effort.

The pay-off

Today, Federal Express has about 52 300 VDTs connected to its mainframes. In addition, 3000–4000 stand-alone PCs are on desktops. This translates into a ratio of about one terminal for each of Federal's 56 000 employees. The company's decade of experience in using VDTs and managing their positive and negative potentials demonstrates that the creation of harmful physical or mental effects is not inevitable from intensive and extensive VDT use — even in a 'pressure cooker' work environment.

By paying attention to the content and context aspects of jobs, providing an employee-supportive organizational climate, addressing 'fair treatment' both in standards and procedures, and spending the dollars and management time necessary to put tools and workplaces into high-quality shape, Federal Express provides a practical model of adept and successful management of VDT use, even in a cut-throat national and international competitive setting. There have been *no* reported cases of repetitive strain injury at Federal Express, despite both intensive and extensive VDT use at customer service centres and various accounting offices. Less than half a dozen stress-related claims by VDT users have been filed under worker compensation in the past 5 years. As one employee interviewed at her VDT in the customer service centre in Memphis put it recently, 'When you give people the best tools, challenge them to do a superior job, and treat them fairly day by day, you get the real pay-off from technology.' In 1990, Federal Express won the highly prized Malcolm Baldridge National Quality Award, becoming the first American corporation in the service sector to be selected. There is no doubt that Federal's 'people-technology' policy played a central role in winning that coveted honour.

Sources

This case study is based on the author's work with Federal Express from 1982–9 in two parallel capacities: (1) as an independent scholar studying various aspects of employee relations and organizational behaviour and (2) as a consultant working with the company on VDT and 'people–technology' issues.

From these two vantage points, interviews related to VDT matters were conducted with more than 150 executives and managers and 500 employees, both in the Memphis (corporate headquarters) area and in field operations in the Northeastern area. In additon, the author conducted an anonymous survey in 1987 that drew responses related to employee relations and technology-related issues from 578 employees and 483 managers.

I wish to record my appreciation to CEO Fred Smith; President James Barbsdale; Senior V.P. for Human Resources James Perkins; former Managing Director for Human Resources Analysis Charles Hartness; present Managing Director for Human Resources Analysis Steve Rutherford; Managing Director for Safety and Health Fred Rine; and Managing Director for Labor and Employment Law Karen Clayborne. I am indebted to the overall management of Federal Express for having the confidence in their policy processes to allow not only scholarly inquiry but also publication of their experiences in a frank and detailed treatment. If more American companies adopted such an approach, their firms, employees and American society would surely be the better for this openness.

Appendix

FEDERAL EXPRESS

PERSONNEL
POLICY & PROCEDURE MANUAL

TECHNOLOGY-PEOPLE IMPACT POLICY

Page 1 of 2
Date 4/3/86

POLICY It is the policy of Federal Express Corporation to systematically incorporate human factors, ergonomics, and job/task design criteria with the development or modification of electronic technology application.

SCOPE All Federal Express jobs which require the operation of electronic equipment such as VDT's, electronic scanning equipment, word processors, and other electronic devices as defined below.

GUIDELINES

General Work functions and environments that are tied to electronic technology and in particular video-display terminals have the heightened potential for causing visual discomfort and self-reports of job-related stress. This policy is applicable to electronic hardware and software, as well as workstations and job design because the latter are inseparable from the former in the promotion of employee health and well-being.

Workstations Defined Workstations are defined as the physical environment in which employees perform their duties including work surfaces, chairs, floors, light sources, noise, and vehicles. All workstations must comply with the Federal Express video-display terminal standards, and the current knowledge about image quality, work place design, and lighting and reflections.

Electronic Hardware and Software These include video display terminals, keyboards, telephone headsets, audio transmission devices, and the computer programs supporting a work function. These devices and computer programs must be selected or designed in accordance with scientific research documenting the maximization of user productivity as well as user preferences.

Job Design Job design is defined as the structuring of work content and context including tasks, responsibilities, performance standards, and breaks. Jobs should be designed so that employees:

- Experience the work as inherently meaningful,
- Feel personal responsibility for the outcome of the work, and
- Receive, on a regular basis, trustworthy knowledge about the results of their work activities.

Job Design Strategies The primary strategy for achieving a successful job design should involve operations managers and job incumbents in the conceptualization and design of jobs requiring the use of technology. Other strategies that might be used to achieve a successful job design include:

- Combining tasks
- Teaming
- Rotation/Temporary Assignments
- Increasing the individual's control over their work
- Establishing client relationships
- Eliminating unnecessary work
- Increasing communications
- Deregulating the work place
- Forming natural work units

Implementation

When new jobs are created and/or when technology is developed or changed in existing jobs, assistance should be requested to assure that ergonomic, human factors, and job/task design criteria have been met so that the resulting jobs/technologies do not adversely impact the people doing the jobs. Assistance is available on request from the following corporate groups:

- Human Resources Analysis
- Network Services
- Organization Planning
- Personnel Services
- Productivity Improvement
- Properties and Facilities
- Safety, Health and Fire Prevention
- Service Systems
- Systems Engineering and Design

Article in 1985 UPDATE, the newspaper for Federal Express employees

Employees Speak; FedEx Listens

By Kem Kelley

When you sit at a video display terminal for the majority of a workday, the type of chair you use becomes a very important part of your worklife.

That's why the company went straight to the user when choosing a new chair design for Customer Service agents at the Memphis, Somerset, and Sacramento centers.

These locations, the first three FedEx Customer Service centers, were not equipped with adjustable chairs like the newer centers are. Adjustable chairs accommodate employees using the same work station at different shifts.

Customer Service agents from the Memphis center evaluated two ergonomic chairs for a two-week period, rating them on adjustability, comfort, mobility, and stability.

The agents' evaluation counted as 75 percent of the deciding vote. The remaining 25 percent was based on the purchasing, properties, and customer service departments' evaluation of price, delivery requirements, and warranties.

The winning chair (by a landslide) was the Steelcase ConCentrx. It won agents' evaluations and was the strongest in pricing, warranty, and dealer network.

The chair will be used in the Memphis, Sacramento, and Somerset Customer Service centers as well as in future centers.

Use of the Steelcase chair in future centers increases the company's Steelcase volume discount for future building projects.

In addition, a study is under way to retrofit the agents' work stations at the early centers to meet the new standards.

The following document is taken from the Federal Express Corporate Interior Guidelines Manual.

8.1.1

VIDEO DISPLAY TERMINALS (VDTs)

RECOMMENDED STANDARDS

GOAL

It is the goal of Federal Express to ensure the health, well-being, and commitment of those employees whose work environment and function is tied to a sedentary existence and the use of video display terminals (VDTs).

The following principles, standards, and practices should govern the design, installation, and operation of all tasks requiring the use of VDTs. These standards are based on extensive research by a FedEx taskforce representing all appropriate departments. The American National Standard for Human Factors Engineering of Visual Display Workstations should be consulted for a more technical reference.

EQUIPMENT AND WORKSTATION DESIGN

The most important general requirement in designing a good VDT workstation is having as many adjustable features as possible. This flexibility is needed to accommodate the diverse physical characteristics of operators, and to incorporate changes as more studies are performed. Features that should be made adjustable include:

- **Detached Keyboard (A).** The keyboard should be detached from the screen so that it can be located according to the preference of the operator. (See Figure 1.)
- **Keyboard Height (B).** This is usually accommodated by using height-adjustable desks or height-adjustable keyboard support surfaces.
- **Keyboard Slope Angle (C).** Operators must be able to adjust the slope angle of their keyboards easily. The keyboard slope should be between 0 and 25 degrees.
- **Display Screen Height (D).** This is made adjustable usually by means of a hardware feature or an adjustable display support platform.
- **Display Screen Angle (E).** This is especially useful in the elimination of glare and augmenting the value of other antiglare measures such as display screen antireflective treatment and well-planned room lighting. An adjustable screen angle range between -5 and +20 degrees is recommended.
- **Chair Backrest (F).** The backrest should be adjustable to support the operator's lower back (lumbar region).
- **Seat Height (G).** If possible, the operator should be able to make adjustments while sitting.

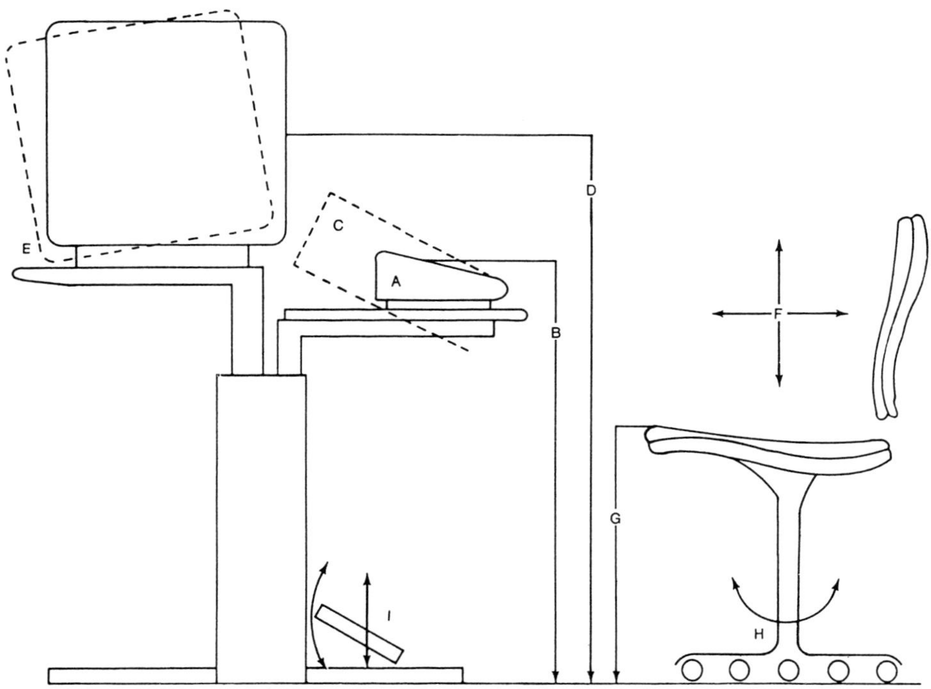

VDT Equipment and Workstation Design Recommendations
Figure 1

- **Other Chair Features (H).** The chair should be movable (five-legged), and the seat should swivel to allow the operator to change position and move easily.
- **Footrests (I).** Footrests are helpful in decreasing stress on the feet and legs, especially if the height adjustments of the supports for display, keyboard, and chair are not sufficiently variable to allow a comfortable posture for every operator who uses a workstation. The footrest should have a large surface and should also have an adjustable inclination and height.

NOTE: Electronic hardware that creates noise (e.g., noise from a blower device in display housing, clicking keys, printer) should be avoided or located away from the operator.

IMAGE QUALITY AND DISPLAY DESIGN

A number of display parameters (e.g., character size, structure, style), image contrast, and stability are known to influence visual performance and operator comfort. For example, operator complaints about the flickering of images are well documented. Techniques are available for measuring parameters that should be considered in the design of displays and for evaluating image quality (see American National Standard for Human Factors Engineering of Visual Display Terminal Workstations).

To ensure proper image quality and display design, the following practices should be required:

- Vendors of display screens should provide the relevant technical information for evaluating image quality. (It is also recommended that equipment be used by operators on a trial basis before acceptance.)
- Routine scheduled maintenance should be provided for all VDTs. Display screens should be replaced when the phosphor begins to wear out and the image sharpness deteriorates.

LIGHTING AND REFLECTIONS

Problems in VDT work places caused by inappropriate lighting can be classified into three general categories:

- Those caused by direct glare.
- Those caused by successive viewing of different luminances (transient adaptation).
- Those caused by reflected glare and veiling reflections.

The following practices are recommended for minimizing these problems.

- In work places with VDTs, an illuminance in the range of 200-500 lux, measured on the work area of the work surface, is normally sufficient. The required illuminance depends on the visual task to be performed. For example, fewer than 200 lux may be desirable for VDT tasks not involving hard copy, while more than 500 lux may be desired for tasks involving the use of poor quality hard copy or high luminance displays. High luminance intensity sources in the peripheral field of view should be avoided. Extreme luminance differences between sequentially and frequently fixated areas, or even only occasionally fixated areas, may produce noticeable performance decrements and discomfort.
- If fluorescent lights are used, they should be fitted with diffusers or parabolic louvers. They should be placed parallel to the sides of the VDT, not directly behind the operator.
- A small task light may be necessary to read source documents in dim room light.
- Problems caused by a bright light source such as a window near a VDT screen can be solved simply by repositioning the screen perpendicular to the plane of the light source or shielding the window with shades.
- Polarized glare covers installed over the VDT screen may be necessary in certain situations to control glare while still allowing operators to have window views. However, because their effectiveness is limited and because they may reduce character image quality, glare covers should be considered only as a supplement, never as a replacement, for control of light through proper lighting design.

The VDT work environment should be arranged so that there are no bright light sources in the operator's field of view. Bright sources in front of an operator can cause contrast glare; bright sources behind can cause mirror-like reflections on the screen. Therefore, the VDT should be oriented so that any bright light sources, such as windows, are to the side.

GENERAL WORKSTATION ENVIRONMENT

A variety of environment factors that affect an operator's health, comfort, and productivity should also be controlled. They are as follows:

- **Noise.** Because noise can be tiring and distracting, the noise level of the room should be kept to a minimum. Carpeting and acoustical ceiling tiles can help to minimize ambient noise.
- **Distractions.** Because the eye, and in particular peripheral vision, is extremely sensitive to movement, the general environment should be free of traffic or any distracting operations. Side partitions on VDT workstations may be necessary to minimize these distractions. However, partitions behind the VDT screen are not advised so that an operator can glance past the screen and relax his eyes. In addition, partitions should not prevent the social interaction of operators, which can be extremely important in maintaining the employee's motivation to perform a routine, sedentary job.
- **Color Schemes.** Color schemes should be neutral and restful to the eye.
- **Ventilation.** Proper ventilation (about two air changes per hour) is necessary to keep operators from becoming drowsy.
- **Temperature.** Temperature and humidity should be kept fairly consistent. Most people are comfortable with temperatures of 68° to 75° F.

Chapter 10

Opportunities for Preventing Mental Health Disturbances with New Office Technologies: Some Problems of Realization

CHRISTOF BAITSCH

Introduction

A widespread consensus exists among work and organizational psychologists that the following set of job characteristics is crucial for the quality of working life.

1. *Autonomy or self-control* — The degree of decision making responsibility.
2. *Feedback* — The degree of getting quantitative and qualitative knowledge of one's own performance.
3. *Task completeness* — The scope or size of a job.
4. Variety — The degree of alternation between different tasks within one job.
5. *Significance* — The importance of a job to the working individual (e.g. Hackman and Oldham, 1976; Emery and Thorsrud, 1976; Ulich *et al.*, 1988).

A closer look at a number of empirical studies reveals that despite the theoretical and methodological problems and differences, their findings demonstrate a similar tendency: restrictiveness in the mentioned job characteristics correlates positively with reduced mental health (e.g. Caplan *et al.*, 1975; French *et al.*, 1982; Ulich and Baitsch, 1987).

Furthermore, we proceed from two suppositions. The first concerns a psychological subject: workers' control should not be maximized in their daily work situation exclusively; control should also extend to the change and design of the work situation itself. Control in the sense of 'having a voice' in the process of design or redesign prevents mental disturbances; under such circumstances where workers are able to anticipate and influence their own future, feelings of helplessness will not occur (e.g. Gardell, 1971; Seligman, 1975; Karasek, 1979). Therefore, we try to meet this human need for control over the immediate environment by a participative approach in work redesign processes.

The second supposition concerns the possibilities for job design, which are opened by new information technology. They promote organizational choice (in the sense of Trist *et al.*, 1963) at shop-floor level as well as in office work. Beside design decisions concerning the human–computer interface, with the new generation of computers the degrees of freedom in designing of organizational structures have become greater; there are now few technical constraints for specific organizational structures. Computer tools may be used for the purpose of increasing or stabilizing centralization and bureaucratization, but they may be used to maintain or regain decentralized structures as well. On the level of job design, this means that computer-aided information technology allows the design of very diverse working conditions (see, e.g. Davis and Taylor, 1976; Kieser and Kubicek, 1977; Taylor, 1979).

To capitalize on these possibilities, it is necessary to widen the scope of analysis and design. Consideration of the single workplace and the single worker must be supplemented with the analysis of the interdependencies of their tasks (Udris and Ulich, 1987). Following the socio-technical system approach, this is necessary, because often it is the combination or division of tasks, which defines the quality of work. In our context this means: (1) The unpleasant effects of VDT work we often observed are, for the most part, avoidable and (2) the quality of working life can even be reinforced if job design is combined with analysing and possibly changing organizational structures. The latter is often the presupposition for success of the first.

Intention of the study

An opportunity to test participative design of VDT work occurred when we were given access to the Personnel Division of a large Swiss chemical company (Troy *et al.*, 1986). In the past the tasks of the Personnel Division were supported by a batch process computer system. In May 1981 the EDP-Division was instructed to develop a new personnel information system (PIS). The on-line system to be developed was intended to contain all the basic functions of the Personnel Division. Improvements compared to the existing system were expected with regard to the data being more up-to-date and of improved quality as well as possible savings from productivity increases. For 45 000 employees a total of 120 pertinent events in 80 applications were to be developed. About 100 individuals within the Personnel Division and some workers in the different plants of the company were affected. By autumn 1983 decisions concerning the structure of the planned system and the hardware to be used were finalized; in addition, a specific software package had been made. At this stage of system development one of the managers within the division succeeded with his proposition to use the technical changes for an organizational restructuring based on a socio-technical approach. For this purpose participation of the end users was a central part and should guarantee that user needs were taken into account in the designing of jobs and organizational structures.

The research setting

In this company, personnel affairs are handled in a centralized manner. The two main departments of the Personnel Division, and those mainly affected by the planned PIS, were (1) the Department for Personnel Administration and (2) the Department for

Personnel Recruitment. The Department for Personnel Administration showed a very strong division of labour. There existed two subdepartments with eight employees each; one subdepartment dealt with all the personnel data, the other dealt with data concerning salaries and wages. Within these subdepartments, tasks were allocated to groups: in the subdepartment 'Personnel Data' work was allocated to two groups, each containing three individuals. One group dealt with attendance data only, and the second dealt with demographic and educational data. Even within these groups there existed specialization, which was very seldom violated.

The resulting tasks can be characterized as follows: data concerning, e.g. attendance at work, are accepted from external stations, registered and changed; as a rule, changing means the encoding of information. Single employees had to process information of a certain type for a certain group of workers, e.g. data concerning shift work or overtime work. Each employee was responsible for 12 000 blue collar workers. The subdepartment 'Salaries and Wages' was organized based on the same scheme. Co-operation among the employees within one subdepartment was not demanded. However, co-operation between employees of the two subdepartments was required, for example, when reports of overtime led to a change of payment. At the level of the individual employee, as well as at the level of the subdepartments, no unity of product and organization existed.

In the other department, 'Personnel Recruitment', organization of work was different. Each of the 15 personnel offices (consisting of a personnel manager and a secretary) had, as a team, responsibility for public relations, recruitment, preselection, contract negotiation and so on, for a certain group of employees, e.g. for the technical staff or for chemical craftsmen.

Secretaries — all female with 3 years of technical training — did not co-operate with each other. This was not necessary because a single secretary, in co-ordination with her supervising personnel manager, performed extensive and complete tasks for a limited number of a specific and well-defined group of employees. Up until the new changes were implemented, the secretaries had sent the data generated by recruitment to Personnel Administration. It was planned that with the new PIS they would enter the data themselves.

Table 10.1. Characteristics of work organization in two personnal departments.

	Personnel Administration	Personnel Recruitment
(1) Relative independency of the organizational units	–[a]	+[b]
(2) Unity of product and organization	–	+
(3) Interdependent tasks within an organizational unit	–	Two-person-group + Subdepartment –
(4) Possibility for self-regulation	–	Two-person-group + Subdepartment –
(5) Boundary regulation through supervisor	–	+

[a] –, Criterion not fulfilled.
[b] +, Criterion fulfilled.

The characteristics of the two departments with respect to the criteria of socio-technical system design can be seen in Table 10.1. The data in this table are derived from the socio-technical system analysis described in the next section.

Project design, methods and research partners

According to the purpose of the study and the assumptions about participation, we chose a combination of classic analytic activities and theory-guided interventions (action research).

1. A socio-technically oriented analysis of the two departments containing mainly the following steps: (a) identification of the system's primary tasks, (b) for the technical subsystem: identification of the work units, their inputs, transformation processes and outputs, types and sources of variances and their interrelationships, (c) for the social system: organizational structure and main groupings, information flow and matrix of variance control, work roles and their interrelationships as well as employees' perceptions of their roles (cf. Emery, 1959).
2. A standardized questionnaire was given twice to all employees of the two departments ($n = 31$); the first before the technical changes (response from $n = 31$) and the second afterwards (response from $n = 27$). Questions concerned demands and satisfaction as well as the evaluations of variety, task completeness, feedback and significance of the job.
3. Based on experts' interviews, four typical workstations in each department were selected and observed each for 1 day; special attention was paid to their interdependencies with other workstations.

Table 10.2 gives an overview over the activities of analysis and intervention. Some characteristics of our sample are shown in Table 10.3.

Results

The observations of eight employees' workdays revealed different results for the two departments. In Personnel Administration all the tasks of a single employee were characterized by identity in content and structure and low demands of training. On the other hand, tasks of Personnel Recruitment employees showed much more variability, immediate feedback and required more training.

Employees' evaluations in the questionnaire and in the group discussions mostly confirmed these findings. Differences in the levels of aspiration between the employees of both departments became visible. The complaints of too little social support were later specified as 'work doesn't allow social support'.

Table 10.4 shows the essential evaluations.

After a feedback of these data to the employees, several group discussions moderated by a member of the research team were arranged. Topics were the meaning of the findings for the employees and possibilities for improving the

Table 10.2. Timetable of analysis and interventions.

Date	Event	Personnel Administration	Personnel Recruitment
	Analysis		
February 1984	Socio-technical system analysis	Work units	Work units
	Task analysis/observations	4 workstations	4 workstations
March 1984	Questionnaire concerning demands, satisfaction, evaluation of, for example, transparency, co-operation	$n = 16$	$n = 15$
	Interventions		
June 1984	Feedback of data concerning observation and inquiry	Subdepartment's groups	Subdepartment's groups
	Group discussions concerning interpretation and correction of findings	Subdepartment's groups	Subdepartment's groups
July 1984	Group discussions concerning different design options for the planned PIS	Subdepartment's groups	Subdepartment's groups
May 1985	*Implementation of the new PIS*		
	Analysis		
November 1985	Questionnaire (identical to questionnaire of March 1984) and questionnaire concerning the perception of changes	$n = 13$	$n = 14$
	Task analysis/observation	4 workstations	4 workstations

Table 10.3. The research partners.

	Personnel Administration	Personnel Recruitment	
		Secretaries	Managers
Number of employees	16	15	15
Women			
unskilled	11	–	–
skilled	1	15	2
Men			
unskilled	–	–	–
skilled	4	–	13
Age (years)			
21–30	3	5	–
31–40	5	7	7
41–50	3	3	5
51–65	5	–	4

Table 10.4. Evaluations of the work situation by the employees in two personnel department.

	Personnel Administration	Personnel Recruitment
Positive aspects of the work situation	Autonomy Relations to supervisors	Autonomy Relations to Supervisors Variability
Negative aspects of the work situation	Too little responsibility Few training demands Too little variability Too little support from colleagues	Too little responsibility Few training demands Few learning opportunities Too little support from colleagues

mentioned negative aspects by means of the planned PIS. The information needed for the discussion of the latter topic was provided by EDP-specialists; also, employees had the opportunity to consult an EDP-technician concerning technical questions that arose.

The result of these meetings in the Personnel Recruitment Department was a consensus to leave the work situation as it was, no structural changes were proposed. Expected time savings should be used for delegation of tasks which were hitherto reserved to personnel managers.

In the Personnel Administration Department, reflections of employees' discussion were aimed at integrating previously separated tasks. Three forms of integration were developed:

1. integration of tasks within the person's own subdepartments,
2. integration of tasks of the subdepartments 'Salaries and Wages' and 'Personnel Data',
3. integration of tasks between the Administration and the Recruitment Department.

Table 10.5 illustrates these propositions beginning with an example from the subdepartment 'Demographic Data'.

It is not simply fortuitous that the older employees especially developed very concrete suggestions concerning the integration of previously separated tasks into tasks of more complexity; they knew such work structures existed from their own experience. Younger employees who had worked predominantly or exclusively under conditions of strict work division expressed discontent with their activity but did not know of any feasible alternatives.

Most of the employees preferred work structures which were equivalent to option 4 in Table 10.5; some also wished to work in structures like option 5.

When these options were presented within the company (spring 1984) the technical system development had already proceeded. Concerning organizational

Table 10.5. *Different solutions for work and organizational structuring of a computer-aided personnel information system.*

	Option 1	Option 2	Option 3	Option 4	Option 5	Option 6
Organizational changes	Structural changes *within* a department (e.g. personnel administration)		Structural changes *within* a department (e.g. personnel administration)		Structural changes overlapping the two departments	
Scope of change	Single work place	Parts of a group	Whole group	Overlapping subdepartments	Integrating personnel recruitment and personnel administration Centralized	Integrating personnel recruitment and personnel administration Decentralized
Changes in work activities	No modification, but more insight to improve clients information	Combination of data, e.g. educational and demographic data Special cases separately as before	Combination of data, e.g. educational and demographic data *and* special cases	Several employees handle together all salary and time matters for a number of clients	Fully integrated personnel authorities in a central department	Fully integrated personnel authorities associated to divisions or factories
Consequences for the employees	Increased insight into more information	Job enlargement Job enrichment possible	Job enrichment Possibly combined with job rotation and/or semi-autonomous groups	Multi-skilling Semi-autonomous groups	Multi-skilling Possibly job rotation and/or semi-autonomous groups	Multi-skilling Semi-autonomous groups Close clients contacts
Consequences for the EDP-System		Suitable to the planned EDP-system without substantial modifications				. . . in connection with local area networks

structures, there was no intention of changing anything; plans were identical to option 1. The renewal was like 'old wine in new skins' as one employee said.

Alluding to the state of system development, the Divisional manager rejected options 3–6; after some discussions management and employees agreed on implementing work structures within the Personnel Administration following option 2.

From a psychological point of view, in May 1985, the new PIS started within unchanged, or only insignificantly changed, organizational structures. Since there was very good training, no problems in operation occurred; 7 months after implementation, the task observation and the inquiry were repeated.

Within the Department of Personnel Recruitment the PIS supported several tasks, but it was one of several tools (see Table 10.7). The VDT was used for about 2 h per day, spread over the whole day. Different tasks were combined to give heterogeneous or mixed work (see Table 10.6), albeit that it depended on the division of work between secretary and personnel manager.

The extent of heterogeneity depended on the extent to which a manager used the VDU himself: if he used it systematically, e.g. for information retrieval, his secretary could use time savings to perform tasks not previously undertaken, especially clerical work.

In Personnel Administration, the new PIS led to more homogeneous work. The VDT dominated most of the tasks, VDT work per day lasted 5 h on average and up to 7 h, for example, at the end of each month.

Table 10.6 shows the distribution of tasks in both departments before and after implementation. In Table 10.7, structures of a typical task in both departments are shown.

Table 10.6. Distribution of different tasks in the departments for personnel administration and personnel recruitment (percentage).

	Personnel Administration		Personnel Recruitment	
	Before	After	Before	After
Secretarial work (typing, filing, etc.)	5–20	2–15	20–45	10–30
Clerical work				
Routine problems	40–70	55–70	5–15	10–20
Special problems	5–12	10–20	15–25	25–40
New problems	0–15	0– 5	5–10	5–10
Total	65–80	75–90	25–40	40–70
Contact with company's employees				
Telephone calls for obtaining information	5–12	2–10	15–25	10–20
Personal contacts for obtaining information	–	–	5–10	5–10
Consulting	–	–	0–15	10–15
Total	5–12	2–10	20–30	25–35
Scheduling work activities	–	–	3– 8	3–10

Source: whole day observations of 2 × 4 workplaces, before and after implementation of the new PIS.

Table 10.7. Typical tasks in the departments for (1) Personnel Recruitment and (2) Personnel Administration.

(1) Administrative handling the 'hiring of an employee' (Department for Personnel Recruitment) Task component	Technical tool	Length of time (min)	(2) Data entry for 23 reports of changed data (Department for Personnel Administration) Task component	Technical tool	Length of time (min)
Data entry with schedules	VDT	9	Data entry with schedules (2 problem cases)[a]	VDT	7
Information obtaining	Telephone	4			
Data entry with schedules	VDT	5	Data entry with schedules (11 cases)	VDT	20
Information obtaining	Telephone	1			
Control of completeness		1	Data entry with schedule (2 problem cases)[a]	VDT	5
Informing another department	Typewriter	2			
Wage calculating	Calculator		Data entry with schedule (8 cases)	VDT	21
	Typewriter	7			
Claiming missing documents	Telephone	1	Printing and checking on entry mistakes		5
Informing training centre	Typewriter	5	Separating, filing and dispatching papers		13
Filing/dispatching documents		5			
Actualizing statistics		3			
Actualizing date book		2			
Technical tools: VDT, telephone, typewriter, calculator, paper/pencil			Technical tools: VDT		
Length of time: 44 min			Length of time: 71 min		

[a] Problem cases are given to the supervisor for solving.

Table 10.8. Comparison of the employees' evaluations of the change of their work situation after implementation of the new PIS.

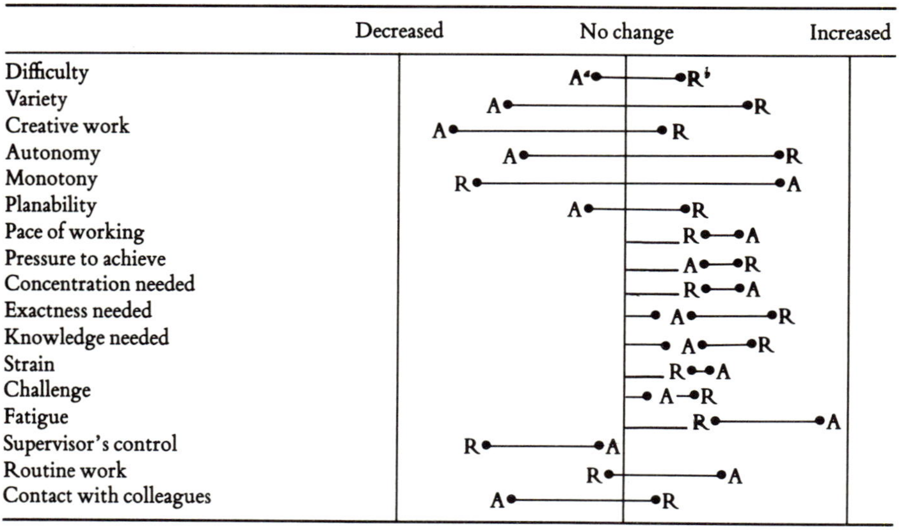

[a]A, Department of Personnel Administration
[b]R, Department of Personnel Recruitment

In the questionnaire the employees evaluated their work situation after using the new PIS for 7 months. The comparison of the employees' evaluation in the two departments shows some trends (see Table 10.8).

The employees in the Department for Personnel Recruitment evaluated their changed work situation as involving increased autonomy, variety, decreased monotony and supervisor's control. This result can be interpreted as a contribution towards preventing mental health disturbances. On the other hand, the employees of the Department of Personnel Administration evaluated their new work situation as involving increased monotony and fatigue, a decrease in variety, creative work and in contact with colleagues, i.e. a decrease in complexity. No contribution towards preventing mental health disturbances can be expected in this department. The opportunity for using the new EDP-system to gain better quality working conditions was not taken in the case of the Department of Personnel Administration.

Twelve months after implementation of the new system in the Department of Personnel Administration there was a dismissal of three employees; in the Department of Personnel Recruitment none of the employees were dismissed.

Discussion

Electronic data processing (EDP) of the 1950s and 1960s is characterized by installations of large computers which in this company caused an increased centralization and an increased division of work in the field of office and administration. The central EDP also led the organization to a centralization of work — a rather immediate effect of technology upon organization. Where technology did not necessitate this outright, technology at least tended to induce it (cf. Taylor, 1979).

We know that EDP-technology does not necessarily lead to a fragmentation of employees' tasks; this is true for the personal information system described in this case study. Here the proposed interactive solutions with EDP allow complex tasks to be left at the individual workplace (see Department for Personnel Recruitment), or brought there (see Department for Personnel Administration). EDP, therefore, can become a tool for employees at their workplace, allowing the performance of tasks which remain as reasonable units or entities, or, as in the case cited, for new tasks to be generated.

Using EDP-technology to design workplaces according to options 4–6 (Table 10.5) is obviously much more appropriate for prevention of mental health disturbances than using EDP-technology to design workplaces as described in options 1–3. If the VDT is only one of the various tools supporting complex task performance, then ergonomic, physical or psychosocial problems are reduced in most cases. Options 4–6 offer opportunities for work design that the employees themselves can adapt to their individual needs and skills (cf. Ulich, 1988).

Probably the most significant danger in this context is that the existing principles of organization stemming from the era of centralized EDP will be continued, even though they are now obsolete.

Reasons for the various technical options not being implemented (as listed below) might be given less frequently by the technicians and specialists in EDP than by those who initiate the projects:

1. the organization of work is scarcely or not at all recognized as possible goals of the projects,
2. the critical decisions in the course of the projects are based predominantly on the interests of EDP, and
3. certain established units of the organization use EDP for their own interests by commissioning or ordering a system that is fitted to the already existing form of organization. This is true especially for the project that is described here.

Within the process of designing new work systems, participation of the end users seems to be especially important; employees on all levels of the hierarchy are willing and able to participate. Additionally, they often find solutions which are otherwise not even discussed, even though they are by no means an expression of employees' interests only, but further better performance and productivity.

Restructuring of work and organization with new technology on the basis only of individual workplaces or isolated organizational units is in danger of achieving less than optimal results; this is true not only with regard to work design that meets human needs, but also in relation to performance and productivity. Instead, a socio-technical approach that includes larger organizational subsystems as the unit of analysis and restructuring efforts is required. The potential of new technology can be realized only when the existing organizational structures are not taken as unchangeable.

This may be summarized in two statements:

1. The introduction of new technology is a unique opportunity for rethinking the company's philosophy and organizational structure.
2. Introducing new technology is not primarily a technical problem, it is a matter of social change and of changing human work activities.

References

Caplan, R. D., Cobb, S., French, J. R. P., Van Harrison, R. and Pinneau, S. R., 1975, *Job Demands and Worker Health: Main Effects and Occupational Differences*. HEW Publication No. (NIOSH) 75–160 (Washington, DC: US Government Printing Office).

Emery, F. E., 1959, *Characteristics of Socio-technical Systems* (London: Tavistock Institute). Doc. 527.

Emery, F. and Thorsrud, E., 1976, *Democracy at Work* (Leiden: Nijhoff).

Davis, L. E. and Taylor, J. C., 1976, Technology, organization and job structure. In: *Handbook of Work, Organization and Society*, edited by R. Dubin, pp. 379–419 (Chicago: Rand McNally).

French, J. R. P., Caplan, R. D. and Van Harrison, R. V., 1982, *The Mechanisms of Job Stress and Strain* (Chichester: Wiley).

Gardell, B., 1971, Alienation and mental health in the modern industrial environment. In: *Society, Stress and Disease*, edited by L. Levi, pp. 148–189 (London: Oxford Press).

Hackman, J. R. and Oldham, G. R., 1976, Motivation through the design of work: test of a theory. *Organizational Behavior and Human Performance*, **16**, 250–279.

Karasek, R. A., 1979, Job demands, job decision, and mental strain: implications for job redesign. *Administrative Science Quarterly*, **24**, 285–308.

Kieser, A. and Kubicek, H., 1977, *Organisation* (Berlin: de Gruyter).

Seligman, M. E. P., 1975, *Helplessness* (San Francisco: Freeman).

Taylor, J. C., 1979, *Fragmented Office Jobs and the Computer*. University of California LA, CQWL-WP-79-1.

Trist, E. L., Higgin, G. W., Murray, H. and Pollock, A. B., 1963, *Organizational Choice* (London: Tavistock).

Troy, N., Baitsch, C. and Katz, C., 1986, *Buerocomputer — Chance fuer die Organisationsgestaltung?* (Zürich: Verlag der Fachvereine).

Udris, I. and Ulich, E., 1987, Organisations- und Technikgestaltung: Prozess- und partizipationsorientierte Arbeitsanalysen. In: *Arbeitsanalyse und Technikanalyse*, edited by K. Sonntag, pp. 49–68 (Cologne: Bachem).

Ulich, E., 1988, Individualisierung und Differentielle Arbeitsgestaltung. In: *Enzyklopaedie der Psychologie, Band Ingenieurpsychologie*, edited by C. Hoyos and B. Zimolong, pp. 511–538 (Göttingen: Hogrefe).

Ulich, E. and Baitsch, C., 1987, Arbeitsstrukturierung. In: *Enzyklopaedie der Psychologie, Band Arbeitspsychologie*, edited by U. Kleinbeck and J. Rutenfranz, pp. 493–531 (Göttingen: Hogrefe).

Ulich, E., Troy, N. and Alioth, A., 1988, Technologie und Organisation. In: *Enzyklopaedie der Psychologie, Band Organisationspsychologie*, edited by E. Roth, pp. 119–141 (Göttingen: Hogrefe).

Chapter 11

Support of a Working Group During Implementation of a New Computer System and Job Redesign in a Life Insurance Company

KARI LINDSTRÖM

Introduction

Aim of the study

Advances in information technology have enabled the insurance sector to develop new, flexible products better meeting individual client needs, and have led to the redesigning of jobs. This chapter describes both the implementation of a new data system for the life insurance sector and an intervention to help employees cope with the difficulties arising once the new data system had been put into use, which involved a gradual shift towards comprehensive, expanded jobs. The redesigning of the jobs occurred simultaneously with the technological change, but was not planned in detail beforehand.

The aim of the study was to help and support the supervisor and other employees of one working group after the new data system had been put into use. The assistance was directed, in particular, to the reorganization of the employees' jobs during this process of change. The consultative roles of the researcher and the occupational health physician were planned so as to take 1 year, and to end with a predominantly qualitative evaluation of the effects of this process of change on job content and the employees' well-being.

Background

This intervention study was a part of an extensive research programme entitled 'New Information Technology and the Work Environment', conducted in Finland during 1985–8. The whole programme dealt with technical, organizational, and safety and health aspects of new information technology and was carried out in three branches: in manufacturing, in the process industry, and in banking and insurance.

Rationales for intervention

The initiative for our intervention effort came primarily from the company's occupational health experts, who wanted to improve their working methods at the working group level instead of concentrating only on individuals. When the research was started in this insurance company, it was decided that a unit of the life insurance department would be the object of study, as the data systems of that department were undergoing major changes. The unit itself to be studied should be one where the employee's average age is rather high. The assumption behind this rationale was that older workers may experience more problems in coping with changes and in mastering new job demands. The final selection of the study groups and initiation of the intervention was, however, left open until the transition to the new data system had taken place.

Job design includes decisions about the tasks to be performed by the workforce, about how tasks are clustered into jobs, and about how jobs are linked together (Davis and Wacker, 1982). It involves both technological and social choices. Job design should also recognize peoples' tendency to form social bonds and help each other (Hackman and Oldham, 1980).

The shift towards comprehensive jobs is a recent general phenomenon in the insurance sector and other service organizations. The new data technology and flexible service products make possible and necessary a new distribution of work. Personnel turnover in insurance and banking is only 2–3%; the low turnover means that in the future the sector will have older personnel facing new job demands. The phases of change are especially critical to problem-solving from points of view of production, the organization, and the individual.

Methods

A case study design and case study methods were used so that the study could be adapted as much as possible to the context in which it was carried out (Westlander, 1985; Yin, 1987). The intervention was based on social support given to the working group during the process of change, and qualitative evaluation of the whole process of change (e.g. Kanawaty, 1980).

Target organization and selection of study groups

The target organization was a Finnish insurance company, founded in 1890 and now having 3680 employees. The company offers life insurance, non-life insurance, reinsurance and pension insurance coverage. Our interest was devoted to the company's life insurance group. The portfolio administration department was selected for closer study.

A working group having major difficulties after the transition to the new data system had taken place, was chosen as the intervention group. This working group, consisting of seven female employees, was responsible for carrying out the changes with respect to the life insurance policies in effect. The work involved mathematical operations concerning payments or decision making concerning the provisions of insurance policies. The employees also informed those insured about the changes in their policies. Six women from the remittance section of the same department served as a reference group.

The median age in the intervention group was 50 years, that of the reference group being 45 years. Most of the employees in both groups had over 20 years of experience at the same company.

Implementation of the new data system

The planning and programming phase of the new data system for life insurance falls outside this study, because it was started 3 years earlier than our study. But it is useful to describe briefly the implementation phase.

The employees at the department were generally informed about the new data system 9 months before it was put into use. The testing of new programs started 6 months before the onset of the new system. Training to use the new data terminals and query applications was given 2·5 months before the onset. The testing of the new system and the training of initiators and on-line users started 1·5 months before the transition. The previous data system for life insurance was taken out of use 2 weeks before the new system was installed. These 2 weeks were devoted to the employees' final training. In the intervention group, it was possible to carry out the non-mathematical provisional changes in life insurance policies just after the transition to the new data system (February 1986), but all changes dealing with money as well as demanding mathematical operations came into use gradually during the next 10 months after the transition. In the reference group, all functions of the new data system operated well from the beginning. That means that the intervention group had an incomplete system, and the reference group a complete system.

Data collection

Data collection was conducted partly by means of two questionnaire surveys, and partly by interviews and group meetings. The time schedule of data collection and study phases are seen in Table 11.1.

Questionnaire survey

The first questionnaire survey, conducted four months (October 1985) prior to the transition to the new data system, included the whole department responsible for changes in life insurance. The second survey was made 20 months (November 1987) after the transition to the new data system. The questions dealt with job demands and stressors, satisfaction with the implementation of the new data system, and subjective symptoms and well-being. In addition, the scale dealing with satisfaction concerning the implementation was presented to both the intervention and reference groups 10 months after the transition (November 1986).

Table 11.1. Time schedule for study and intervention process.

Year/Month	Study and intervention process
1985	
October	Questionnaire survey I
November	
December	
1986	
January	Interview of supervisors
February	*Transition to the new data system*
March	
April	
May	Interview of supervisors
June	Survey feedback
July	
August	
September	Selection of the study groups
October	Information to the company management
November	Interview of supervisors and employees
December	
1987	
January	Feedback of interview
February	
March	
April	Interview of the supervisors
May	Action = ergonomics
June	
July	
August	
September	Action = developmental discussion
October	Action = summary at group level
November	Questionnaire survey II
December	
1988	
January	
February	Group discussion with supervisors and employees = evaluation of the outcome

Interviews of supervisors and employees

The departmental director and section supervisors were interviewed 1 month before the transition to the new data system (January 1986), and again 3 months after transition (May 1986). The intervention part of the study started with the interview of the departmental director and the supervisor of the intervention group (November 1986). The supervisor was later interviewed repeatedly during the study, at about 3–4 month intervals. These interviews dealt with the functioning of the new data system, its effects on the distribution of work, the content and organization of jobs and the employees' well-being and functioning.

The employees in the study groups were interviewed thoroughly 10 months after the transition to the new data system (November 1986). The interviews dealt with their work, satisfaction about the data system, and well-being.

Job Redesign

The jobs of individual employees in the study groups before the transition to the new data system were rather narrow and fragmented. The same life insurance was handled in two separate departments and by several employees. Therefore it was decided to redesign the jobs in connection with the implementation of the new data system. The optimal goal, set before the transition to the new data system, was that any life insurance policy should be handled in the same department and by one employee.

The reorganization was designed to start with reorganizing the various departments and sections, followed by the redesign of idividual jobs. The initiative for the redesign of jobs came from the company management, and was accepted at department and section supervisors' level 2 years before the actual change. Attitudinal preparation to pave the way for job redesign was carried out among the employees when the new data system was being planned. The employees were encouraged to become identified with job design.

The job redesign was accomplished by combining tasks, the aim being jobs characterized by skill variety and task identity (Hackman and Oldham, 1980). Other criteria applied were based mainly on factors affecting the quality of the working life, and were especially related to job content and to internal social relations (Davis and Cherns, 1975). Other actual factors related to job content were feedback, challenge, individual autonomy and self-regulation, opportunity to use skills and capacities and perceived contribution of own work to service product. The factors related to internal social relations in the working group were the opportunity for social contacts, recognition of achievement, the opportunity for mutual social support, and a good working climate. The physical environment was taken into consideration when the new data terminals were introduced.

Structure of intervention

A weekly meeting of the intervention group was introduced at the beginning of the intervention approach (January 1987) and was continued during the intervention.

The researcher and the occupational health personnel acted as consultants on the following topics:

1. *Training and guidance* — Guidance of the supervisor and the working group through the job redesign process, especially providing them with knowledge about criteria concerning the quality of the working life and the length and phases of the process of change.
2. *Process facilitation* — (a) Helping the working group in all phases of the process when obstacles were encountered and when their own confidence in the goals was low, and (b) giving feedback during the process of change and relating it to the process at company level.

3. *Intermediary link* — Giving neutral information about the nature of the process of change from various points of view, e.g. the perspectives of the departmental director, the supervisor, the individual employee and the data processing expert.
4. *Research* — (a) Assessment of the effects of job redesign on job demands, and (b) collection and analysis of data on well-being, and presentation of the findings to the working group at various phases.
5. *Agent of change* — At the organizational level, to give information about the progress and positive reinforcement regarding successful trials.

A separate part of the intervention was to integrate expertise within the company so as to support and promote the job redesign. These efforts were directed at the physical working conditions (i.e. ergonomics) and factors of job content (i.e. individual development discussions).

Ergonomics

Ergonomic problems were discussed in the weekly meeting of the intervention group (May 1987), partly on the basis of the results of the questionnaire survey done at the beginning of the study (October 1985). All members of the working group, including the supervisor, and the occupational physician, the physiotherapist (ergonomist) and the researcher attended the meeting. The problems discussed dealt with lighting conditions in computer terminal work and constrained work postures in sitting work, and problems involving neck, shoulder and low back pain.

During computer terminal work, the glare of sunlight coming in from the windows caused definite symptoms of eye irritation. In practice, it was improved easily by adequate venetian blinds. Earlier attempts to get venetian blinds had not succeeded because the offices were scheduled to be moved to another building in 2 years. Responsibility for solving the lighting problems rested with the occupational health service personnel.

Overtime work and stress caused by the process of change increased problems involving the neck, shoulders and low back. Exercises during breaks at the workplace had ceased once the work piled up. The physiotherapist tried to motivate the employees to start doing exercises during breaks again, and to do other muscle relaxation.

Individual developmental discussion promoting the redesign of jobs

The individual development discussions introduced by the company's personnel development unit were integrated to promote the redesign of jobs (September–October 1987). These discussions dealt with the training needs for the coming year and with the employees' long-term developmental needs and goals. It was also possible to give and obtain feedback about supervisory practices. The co-ordination of individual and working groups' goals were also checked.

A series of concrete questions and discussion topics were proposed for these discussions on the basis of the research results. These covered the problems noted during the process of change in the organization, problems involving the content of

Table 11.2. Topics in the questionnaire for developmental discussion between supervisor and employee.

1. List your main tasks (duties) and evaluate which of them would need more or less time than right now.
2. Which of your duties would you want to get rid of and why? Is the task totally unnecessary, or who could it be transferred to?
3. Has someone else got tasks that should be transferred to you? What are these tasks and where are they being done now?
4. What kind of problems do you see in the distribution of tasks within your working group or in co-operation with other working groups? What kind of solutions would be appropriate?
5. Do you see anything wrong with how your own working group functions? Has someone got too much to do, and someone else too little to do? How would you solve these problems?
6. Who takes care of your tasks during your vacations, sick leaves or other absences? How would your tasks be carried out best during your absence?
7. What kind of tasks are piling up, when do they pile up, and what can be done to avoid such situations?
8. Have you got any skills or knowledge you cannot use continuously in your present job?
9. What kind of training do you and your co-workers need? Give the topic and who needs training in it.
10. What kind of guidance, help or support do you need from your supervisor?
11. What is the most disturbing or causes insecurity or difficulties in the functioning of your working group? How would you propose to make the situation better?

work, and problems relevant to the employees' qualification and well-being. A special questionnaire to be completed (September 1987) before the discussion included the topics and questions shown in Table 11.2.

After the individual discussion, the working group as a whole set the goals for development at the group level (October 1987). One group meeting was devoted to joint planning of ways in which the working group could reach the set goals.

Results

Satisfaction with the implementation of the new data system

Ten months (November 1986) after the transition to the new data system, the employees first evaluated their satisfaction with advance information, training and guidance, the role of data processing experts and supervisors, the consideration of human issues and the amount of time and personnel.

The evaluations of the reference group were more positive than those of the intervention group. This finding was expected, because the transition to the new data system had gone well for the referents, whereas the study group had experienced various delays and errors. During the first 6 months after the transition, the life insurance documents piled up, mostly because of the incompleteness of the data system. The delays resulted partly from the low priority of this component of the system, as it was not very important to direct customer service.

Participation in the planning process of the company was previously almost non-existent at the ordinary employee or end user level. When the new system was planned, keeping the employees informed was not overlooked and at least some of the employees to be affected by the change participated in its planning.

Advance training was evaluated as being somewhat unsatisfactory by the intervention group, but was well compensated for by the easily accessible help and guidance within the group.

The intervention group evaluated the data processing experts' knowledge of their own work content as very poor. This poor rating was probably caused by major delays encountered with some of the data system needed by the intervention group, and by the many programming errors disturbing and even preventing their work.

Both groups, but especially the intervention group, evaluated the amount of personnel and the time reserved for becoming acquainted with the new system as insufficient.

After 20 months of experience with the new data system (November 1987), the same aspects of implementation were evaluated again. In general, the employees in both groups, but especially in the intervention group, were now more satisfied than 10 months earlier. Dissatisfaction with the data processing experts' knowledge about the employees' work had decreased in the intervention group. Lack of personnel was still considered to be a problem in both groups.

Interviews and group discussions: effect of new data system and job redesign on work content, social contacts and well-being

Work content

The goal in the company was to reorganize the work by combining tasks, because the traditional way of organizing work was quite fragmented. Implementation of the new data system enabled this.

In the intervention group, by the end of 1987, the employees changing the life insurance policies were able to handle the job from beginning to end and not only the mathematical operation phase. They also recorded data into the data system, checked the information letter printed out, and mailed it to the client. The data input phase in the new system was experienced as being difficult. Those who made changes in the provision of insurance policies also took care of the whole work process from the beginning to informing the client. However, in practice many problems with the functioning of the new data system occurred in the intervention group during the first 18 months after the transition to the new system. The idea that everyone can do everybody's work, however, did not prove to be a fully realistic goal. Although work piled up during the transition phase, no return to the earlier fragmented tasks occurred in this group, as was the case in some other departments.

During the period when work was piling up, no reorganizing of the work, i.e. no other efforts to combine tasks was made other than that required by the new data system, but 18 months after the onset of the new system, some employees gradually received further taining in order to expand their jobs. It was apparent from the beginning that no general changes in task structure could be made, but that changes should be individually oriented. The issue of substitutes for co-workers on vacation or sick leave was still unresolved 10 months after the onset of the new system.

The biggest problem in organizing work at that time was how to solve the fact that work was piling up. The testing of new parts of the data system was still going on simultaneously, and required nearly the full-time attention of one of the seven employees.

As to the reference group, the new data system began to function without major difficulties after the transition (February 1986). In interviews carried out after 10 months (November 1986), it was perceived as making the work easier. It became quicker to check errors, and the amount of manual tasks decreased.

The idea of comprehensive, combined tasks was still viable 10 months after the actual transition in the reference group, but the attitudes towards it were somewhat divided. About one half of the personnel wanted more comprehensive tasks, when the other half was against it. This was probably partly due to poor attitudinal preparation within this group.

Social contacts and well-being

The intervention group had a good working climate, with a new supervisor being appointed just before the implementation. In spite of difficulties with the data system, the supervisor's supportive role, and good working climate in general, were of great importance. The reference group did not have their own supervisor during the first year after the transition, which caused some problems in co-operation.

The employees in the intervention group felt that they got the necessary information, help and emotional support from their co-workers and supervisor. Their work motivation was high, and they were task-oriented. All had long work experience within the insurance company, mainly at their present jobs. However, the problems caused by the incomplete and erroneous data system had caused increased fatigue and lack of satisfaction with their own efforts. The intervention group also perceived a lack of esteem from the company. In the reference group the occupational qualifications and work motivation also seemed to be good.

The discussions in the intervention group at the end of the follow-up period (at the turn of 1988) showed that after the period when work was piling up, it was possible to return to normal working hours and pace. Excess fatigue, eyestrain, elevated neck and shoulder pain and dissatisfaction with one's own efforts were no longer sources of complaint and they felt themselves a sense of relief. Adverse physical symptoms were not a source of complaint with the reference group employees during interviews (November 1986) or on the questionnaire at the end of study (November 1987).

Questionnaire surveys: the effects of the process of change and intervention on job demands and well-being

These results are based on comparisons between two similar questionnaire surveys conducted in October 1985 and November 1987.

Table 3. Perceived effects of the new data system and job redesign on tasks, job demands and well-being in November 1987.

	Intervention group			Reference group		
	Decrease	Same	Increase	Decrease	Same	Increase
Tasks						
Typing	+			+		
Manual calculations	+			+		
Mental work	+				+	
Inspection	+				+	
Instructions to be taken into account		+			+	
Social contacts						
Contacts with co-workers		+			+	
Job demands						
Individual autonomy		+			+	
Work pace			+			+
Use of one's own abilities		+			+	
Variety of job content		+			+	
Challenge		+				+
Job appreciation		+			+	
Qualification demands		+			+	
Well-being						
Feeling of achievement			+		+	
Sense of mastery of the work			+		+	

The effects of both technological and other changes on one's own work situation were evaluated as being more marked by the employees in the intervention group than among the reference group (November 1987). The intervention group also evaluated these effects on their work as being entirely positive, whereas in the reference group the effects were evaluated in various ways, ranging from positive to negative.

The evaluations of the job content variety before and after the change were about the same. In both groups, jobs were evaluated to be quite varying, and no one perceived the work as monotonous. After the 20-month follow-up period (November 1987) in the intervention group, more employees perceived their job as very varying as compared to the situation before using the new data system and before the other changes.

The frequency of manual routine tasks, such as typing and calculations, had decreased in both groups, but amount of mental work and the number of inspection tasks decreased only in the intervention group. The reference group found their work to be more challenging at the end of the study period (November 1987), whereas the feeling of achievement and mastery of one's own work was on a higher level in the intervention group. In both groups the work pace was experienced as having increased during this follow-up period (Table 11.3).

No one in the intervention group complained of continuous or frequent excessive fatigue, headache, nervousness or sleeping difficulties on the questionnaire survey 20 months after the transition (November 1987). These symptoms were at the same level as before the whole process of change. Eyestrain was somewhat more frequent after the follow-up period than at the beginning, but the continuous aches and pains in the neck and shoulders that had been reported to be slightly elevated at the beginning and clearly elevated during the follow-up period had nearly disappeared by the end of it. In the reference group these symptoms were at about the same level throughout, and at the time of questionnaire survey II (November 1987), there were no marked differences between the study groups.

Discussion

Summary of approach and main findings

This qualitative case study was meant to follow the changes in a work unit where a new data system and redesign of jobs by combining various tasks were being implemented. The study included interventions to help and support the working group cope with difficulties associated with the implementation of the new data system and the reorganization and redesign of individual jobs.

The role of the researcher was consultative. The company's occupational health service personnel participated in the study in the same way, following the process of change and participating in supportive interventions. Another motivation for their participation was to improve their own knowledge and working methods at the group level.

A working group in which the implementation introduced major problems was chosen to be the object of intervention. The delay and the great number of errors in the new data system in the intervention group was unexpected, but served as an extreme example of how employees are capable of coping with external obstacles when they have good professional skills and a mutual social support system.

Ergonomics was chosen as the starting point in the intervention programme because it was a concrete and traditional method used by occupational health personnel. The working conditions should also be optimal before actions aimed at changes in work content can be meaningful.

The feedback and group discussions revealed that the piling up of work was the main problem during the first year after the transition. All group efforts were thus directed to its solution. The active developmental and job redesign activities could be continued only when the work backlog was cleared some 16 months after the transition. These intentions, however, were kept alive during the difficult period when work was piling up, i.e. at a time when the old, fragmented working method would probably have been more effective.

At the end of study period, the intervention group had overcome the excessive fatigue and elevated somatic strain symptoms related to problems during the transition. The ergonomic intervention was also meant to remedy those. The intervention group had better mastery of their work and better quality of work results at the end of the study when compared to the reference group.

Importance of the findings

The main goal set for the intervention was to support the idea of comprehensive insurance work. The quality of working life criteria, especially those of, for example, job content variety and task identity, were supported and promoted during the intervention.

The evaluations of the intervention group were more positive concerning their jobs in general at the end of study period. In the reference group, the main problem was the lack of a supervisor, and they were more polarized in their attitudes about job redesign. Both groups evaluated more positively the outcome of change than did employees in another department, where the comprehensive insurance work was first implemented but then cancelled because of the backlog of work created after the transition to the new data system. The supervisor of the intervention group appreciated that the system was more effective because it demanded less manual work and because it was easier to check errors when it was known who was responsible for the handling of a single insurance document.

The questionnaire survey at the end of the study showed that the work of both groups, but especially of the intervention group, contained less manual routines. However, the work pace was perceived as having increased. The reference group found their work to be more challenging than did the intervention group, but it is difficult to know whether this depended only on the novelty of the new working methods. In the intervention group, the feelings of achievement and mastery of the work were reported to have increased, but not in the reference group. The use of one's own abilities was perceived as being about the same. This was quite reasonable, as the work content variety had already been rather high, even before the change.

The overall conclusion for the intervention was that when the groundwork is laid for positive attitudes to the implementation of a new data system and job redesign, the base for a good outcome, in terms of job demands and the employees' well-being, is mutual social support and access to outside support, especially during difficult phases of the process of change.

Limitations of the study

Because the study was conducted with a case study design and the appropriate case study methods, only qualitative evaluations were possible. However, these were supplemented with quantitative methods. Comparison of the two groups gave some basis for conclusions on the effects of intervention, but the situation of the study groups was rather different in many respects. The most important difference was that the reference group had no supervisor of their own at that time, the department director acting as their supervisor, too. The situation within the department for both groups was similar as to support from the departmental director, who kept the idea of redesigning jobs alive. The situation was the opposite in a parallel department.

The data system implemented was the same for both study groups, but the implementation caused more difficulties and problems for the intervention group. In spite of this, however, the outcome was better among the intervention group. The main difference favouring the intervention group was the primarily better mutual social support and extra support derived from the researcher and the occupational health personnel.

The effects of the consultative role of the researcher and the occupational health personnel on the outcomes of health and the quality of work results cannot be separated out because so many other factors differed between the study groups and their situations.

Acknowledgements

I thank the Finnish Work Environment Fund for financing this study as a part of the research programme 'New information technology and work environment'.

References

Davis, L. E. and Cherns, A. B., 1975, *The Quality of Working Life*, Vols 1 and 2 (New York: Free Press).

Davis, L. E. and Wacker, G.J., 1982, Job design, In: *Handbook of Industrial Engineering*, edited by G. Salvendy, pp. 2.5.1.–2.5.31 (New York: John Wiley).

Hackman, J. R. and Oldham, G. R., 1980, *Work Redesign* (Reading: Addison Wesley).

Kanawaty, G., 1980, Introducing and developing new forms of work organisation. In: *Managing and Developing New Forms of Work Organisation*, edited by E. Kanawaty, pp. 169–206. Management Development Series No. 16 (Geneva: International Labour Office).

Westlander, G., 1985, *Om contextorienterad ansats. En metoddiskussion med anknytning till organisationspsykologisk tillämpning*. Undersökningsrapport 8. (Stockholm: Arbetarskyddsstyrelsen).

Yin, R. K., 1987, *Case Study Research*. Applied Social Research Methods Series, Vol. 5 (Beverly Hills: Sage Publications).

PART III
Clinical evaluation and management of health complaints

Chapter 12

Vision and the Video Display Terminal: Clinical Findings

JAMES E. SHEEDY AND STEVEN D. PARSONS

Introduction

Visual problems among users of video display terminals (VDTs) have been an unfortunate by-product of this advancing technology. Eyestrain, visual fatigue, blurred and double vision, as well as headaches have been reported. The proliferation of VDTs in the workplace and the prevalence of these problems has resulted in widespread study into the nature of these complaints. Matula (1981) compiled a list of 174 references pertaining to the ocular effects of VDT usage. The widely dispersed literature on the effects of VDT usage on the eyes has touched upon many aspects of the problem, however, it is still not clear whether there is a higher incidence of visual complaints among VDT users compared to other office workers. Smith *et al.* (1981) found that VDT users reported a higher level of visual complaints than non-VDT users and suggested that the solution might be related to job and workplace redesign considerations. Padmos (1987) found visual fatigue to be the most frequent health complaint with other types of work and not exclusive to VDT use. Howarth and Istance (1985) found equal complaints among users of VDTs and users of hard copies. Other studies have been divided in determining whether VDT use is, in fact, a significant causal factor for increased visual problems.

Whether or not usage of VDTs results in greater symptomatology than other office tasks, *the reasons for the symptoms still must be identified and rectified.*

The most likely causative factors can be categorized as visual or environmental in nature. It is extremely unlikely that any single factor is causing the discomfort for all users. For any given user, however, there may be a single cause for the discomfort or a combination of causes. Visual causative factors refer to specific diagnosable visual disorders which an individual may have, whereas environmental causative factors refer to specific factors in the design of the VDT, the workstation and the room environment.

This chapter reports on the findings of a clinic designed to identify and treat or rectify the causative factors of the symptomatology. This clinic, established in August 1985 at the University of California, Berkeley, School of Optometry, was designed to relieve the symptoms of VDT users by diagnosing and treating visual problems and by making suggestions for ergonomic improvement of the individual patient's workstation.

After making an appointment with the clinic, the patient was sent a questionnaire which served a two-fold purpose. First, the patient was asked to critically evaluate the types of symptoms they were experiencing such as headaches, blurred vision, double vision and eyestrain. The patient was also asked to evaluate the severity of the symptoms. Second, the patient was asked about their workstation environment and to make measurements of that environment. This involved questions and measurements concerning lighting, reflections, table height, chair height, keyboard height, eye to screen viewing distance, location of reference material and screen tilt. These observations and measurements were used to create a simulated work environment in the clinic.

Each examination began with a complete primary care eye examination which included an assessment of refractive error, binocular vision, the internal and external health of the eyes, visual fields and intraocular pressure. When indicated, treatment, referral or photo documentation was made. Secondary level testing of accommodative (eye focusing) abilities, binocular vision, eye movements and glare sensitivity were also performed.

After the eye examination the patient was seated in a VDT workstation simulation. The workstation and screen colour were arranged to match the observations and measurements which the patient had previously provided on the questionnaire. This station could serve several functions, depending upon the needs of the patient. This could include demonstration of any spectacle correction being considered or recommended. The patient could observe the effects of the optical correction and help make refinements in the correction while in the simulated work environment. Effective decisions could also be made concerning single vision lenses versus bifocal lenses, type of multifocal lens correction, and location of the bifocal segment within the lens. The effect of lens tints upon screen display visibility could be demonstrated. Numerous auxiliary items such as anti-reflection filters, document holders, localized lighting, wrist support, feet support, and so on, could be demonstrated and suggested if appropriate. The workstation design, arrangement and lighting were discussed with the patient and suggestions for improvement were made.

Six weeks following the examination, a follow-up questionnaire was sent to each patient seen in the clinic. This post-examination questionnaire asked what recommendation(s) was(were) made, the patient's compliance, and the degree of relief from symptoms. If the patient was still experiencing symptoms, their type, frequency and severity were determined with the same set of questions as on the initial questionnaire.

The VDT examination was designed as a primary care examination with a special emphasis on the needs of the VDT user.

Methodology

The patient charts of those patients examined in the VDT eye clinic between autumn 1985 and spring 1987 were retrospectively analysed. The following data were obtained from the 153 complete charts that were analysed:

1. age,
2. sex,
3. hours worked per day,
4. months of VDT usage,
5. constant versus intermittent use,
6. pre-examination symptoms (as listed on the pre-questionnaire),
7. pre-examination appraisal of the severity and frequency of symptoms,
8. diagnoses, and
9. treatments.

From the 56 patients who had returned the post-questionnaire, the following data were obtained:

1. post-treatment symptoms, if any,
2. post-treatment appraisal of the severity and frequency of symptom(s),
3. patient rating of the degree of symptom relief after treatment at the VDT clinic.

On the pre- and post-examination questionnaires, each patient was asked to categorize the symptoms as to: (1) type of complaint (headaches, eyestrain, blurred vision, diplopia, or others), (2) severity (just noticeable, tolerable, very bothersome, or must seek immediate relief), and (3) frequency (monthly, weekly, every other day, or daily). Symptom severity and frequency were each scored from 1 to 4 depending upon the response (e.g. monthly = 1, very bothersome = 3, etc.). The post-questionnaire additionally asked the patient whether they had complied with the recommendations and to rate their symptomatic relief from the following: complete relief, mostly relieved, slightly relieved, no change, and symptoms worse. A paired *t*-test was performed on the severity and frequency scores, as well as their sum, comparing the pre- and post-scores to determine if any significant change occurred in symptom severity and/or frequency as a result of examination and treatment.

General characteristics of the patient population were derived, including the average age, number of males and females, average hours per day usage, average months of VDT usage, and number of constant and intermittent users. Additionally, a tally was made of the numbers of symptom types and the various diagnoses and treatments.

Chi-square calculations were performed to compare each diagnosis and treatment with each symptom and level of symptom relief across the population. This was to determine if there was a meaningful correlation between a given diagnosis or treatment with a specific symptom or level of symptomatic relief after treatment. A total of 255 chi-square calculations were made.

Results

General

As shown in Table 12.1, the average age of a patient visiting the VDT Eye Clinic was 38, with most patients between 28 and 48 years old. Patients worked an average of 6 h per day on the VDT but most considered themselves to be intermittent rather than constant users. The average patient had used the VDT for more than 51 months, but there was a lot of variability in this measure indicating that both long-term and new users visited the clinic. Almost twice as many women compared to men visited the clinic.

Table 12.1. Characteristics of the sample.

Age (years) $x = 38\cdot30$, range: 22–71, s.d. + 10·45
Sex: 53 males, 100 females
Hours per day: $x = 6\cdot02$, range: 1–11, s.d. = 2·25
Months usage: $x = 51\cdot29$, range: 1–180, s.d. + 40·41
Constant/intermittent: 64 constant, 80 intermittent

Visual

Tables 12.2 and 12.3 document the visual diagnoses and treatments given in the clinic. These diagnoses and treatments were divided into primary, secondary and tertiary categories. The primary diagnosis/treatment was clinically judged as the major contributing factor to symptoms. However, for many patients there were other contributing factors which were categorized as either secondary or tertiary.

Accommodative problems were the most frequently diagnosed. Accommodation refers to the ability of the eye to change its focus for different working distances. There are two main clinical components to accommodation: amplitude and facility. Amplitude refers to the dioptric strength of the focusing mechanism, i.e. how close to the eye one can maintain a clear image. Facility refers to the flexibility of the focusing mechanism, i.e. how rapidly and efficiently one can change focus from one distance to another. Deficits of either facility or amplitude of accommodation can cause performance problems. Intermittent blur, distance blur after near work, and momentary blur when looking from distance to near can be the results of poor facility of accommodation. Near blur can result from decreased accommodative amplitude. These problems were very frequent in our VDT population (Table 12.2) and appear to be more common in this population than in other clinical populations; however, data from other populations are not available.

Accommodative disorders were diagnosed in the VDT Eye Clinic by clinical judgment of the monocular and binocular amplitudes of accommodation, plus and minus binocular lenses to blur at 40 cm (NRA/PRA), and binocular and monocular

Table 12.2. Number of patients diagnosed with each condition — categorized by primary, secondary, or tertiary diagnosis.

VDT diagnoses	1°	2°	3°
Poor facility of accommodation	25	6	0
Uncorrected presbyopia	21	2	0
Advancing presbyopia	16	0	1
Improper spectacle design	15	4	0
Esophoria	12	6	1
Reduced amplitude of accommodation for age	8	2	0
Exophoria	8	6	1
Astigmatism	7	4	1
Anisometropia	5	3	0
Reduced vergences	5	5	1
Hyperopia	4	11	0
Dry eye syndrome	4	0	0
Contact lens problems	3	4	0
Current prescription overminused	2	1	0
Strabismus	1	1	1
Myopia	0	1	1
Hyperopia	0	1	0
Retinal hole	0	0	1

Table 12.3. Number of patients receiving each treatment — categorized by primary, secondary or tertiary treatment.

VDT treatments	1°	2°	3°
Presbyopic single vision prescription	46	3	2
Pre-presbyopic reading prescription	31	1	0
Vision training	22	6	0
Correction of astigmatism	7	2	1
Presbyopic multifocal	6	0	0
Correction of hyperopia	5	6	0
Correction of anisometropia	4	3	0
Regular rest breaks	3	2	0
Correction of myopia	3	0	0
Correction prescription given	2	1	0
Artificial tears	2	0	0
Contact lens refit	2	0	0
Contact lens solution modification	2	0	0
Tint	1	0	1
Increased blink rate	1	0	0
Pre-presbyopic multifocal	1	0	0
Prisms	0	4	0
Polish contact lenses	0	1	0

flipper rates (+/−1·50 D). When these measurements revealed an accommodative dysfunction, the most common treatment for this problem was a prescription of plus lenses to compensate for the accommodative deficiency (Table 12.3). An alternative treatment was visual training. The treatment of choice for a particular patient depended upon the clinical findings and upon the time availability and preferences of the patient.

Previously uncorrected presbyopia and advancing presbyopia were the second and third most frequent diagnoses (Table 12.2). Presbyopia is the condition in which the normal age-related decline in accommodative amplitude results in an inability to see clearly at near. After about the age of 40, most people need a near prescription to help in near tasks. It is not surprising that presbyopia was found to be a frequent diagnosis, considering that the average age of the population was 38 years old.

The fourth most frequent diagnosis also involved presbyopia — an improper spectacle design (Table 12.2). The common treatment for presbyopia is a bifocal prescription in which the top portion of the lens has the distance prescription and the bottom portion of the lens has the near prescription. Eye doctors most commonly design the near prescription for a 16″ viewing distance and the bifocal segment is located for a downward viewing angle of about 25°. However, the VDT screen is most commonly located farther away (20–24″) and higher (10–15° below straight ahead) in the field of view. The inappropriate prescription and lens design for the VDT task causes the user to tilt their head backward and lean inward towards the VDT, resulting in neck aches, back aches and poor vision. The bifocal segment can be placed higher in the field of view and reduced in power in order to better meet the needs of the VDT user. However, this design precludes the use of these spectacles for other purposes such as walking around the office (because of the high bifocal segment).

As indicated in Table 12.3, the most frequent treatment for presbyopia was a low plus (+0·50 to +0·75) single vision prescription. This prescription and lens design eliminates the difficulties associated with bifocals and provides a wide field of view. This type of prescription will produce a slight distance blur, but this is usually acceptable since office workers seldom look critically at distant objects. Younger presbyopes can still accommodate from the mid-range to closer distances through plus lenses. For advanced presbyopes, bifocals that incorporate the intermediate lens power in the top of the lens and the near power in the bifocal segment work well.

Eye alignment is a contributing factor to the symptomatology associated with VDT usage. When viewing an object at near distances, individuals with binocular vision must keep both eyes aligned on the object of regard. Many patients have difficulty maintaining this alignment. They may have an esophoria (eyes tend to cross or converge) or an exophoria (eyes tend to diverge). Quite a few VDT users were diagnosed to have an esophoria or exophoria which was clinically significant. Esophoria at near often accompanies accommodative disorders and is usually best treated with reading glasses. An exophoria at near is usually associated with convergence insufficiency, the best method of treatment is visual training to enhance the convergence ability.

Hyperopia was diagnosed in several patients as shown in Table 12.2. Hyperopia, or far-sightedness, is a refractive condition of the eye where the image of a distant object focuses behind the retina without any accommodation. To see a near object clearly, such as a VDT screen, a greater accommodative effort is required compared to non-hyperopic individuals. Patients in their twenties and thirties who otherwise might not need a reading prescription if they were working at a less visually demanding task, now benefit from such a correction. This helps the patient to maintain a clear image over a sustained period of near work by decreasing the amount of effort to focus. Hyperopia is treated by prescribing plus lenses to provide clear, comfortable vision at the distance of the VDT.

Astigmatism was diagnosed in several patients. In this refractive condition, the rays of light from an object do not come to a point focus, but are focused as two line images at different distances from the retina. This causes blur at both distance and near. This condition is corrected with a spectacle prescription.

Myopia, or near-sightedness, is a refractive condition which results in clear near vision without needing to use the accommodative mechanism as much as individuals without myopia. Consequently, near work at a VDT requires less focusing effort for the myopia. It is not surprising that myopia was not diagnosed as a causative factor in symptomatology.

VDT users who use contact lenses instead of glasses sometimes experience problems with dry eyes. This occurs when users do not blink as often with sustained viewing as they otherwise would. These types of problems occur more frequently among users who look at the screen continuously instead of looking regularly at other materials while working at the VDT. These individuals can be taught to blink more frequently. Artificial tears may be indicated to relieve dry eye symptoms.

Environmental

Table 12.4 presents the most frequent environmental diagnoses and treatments identified and handled by the clinic.

A major environmental problem involves lighting. The typical VDT office is illuminated with overhead fluorescent light. The illumination level on desk surfaces is often 100–150 ft candles, well above the 18–46 ft candles recommended by the ANSI

Table 12.4. Frequent environmental diagnoses and treatments.

Diagnoses	Treatments
Glare from lighting	Anti-reflection screen
Document location	Document holder
Reflections	Visor
Screen brightness	Adjust screen brightness
Image screen quality	Lighting alterations
Improper screen height	Changes in furniture
Improper keyboard height	Changes in workstation
Improper furniture	Louver screen
Ergonomics	General environmental
Other lighting problems	

Standard (1986). This results in a surrounding environment which is considerably brighter than the dark screen at which the worker is looking. This is often further exacerbated by auxiliary lighting on reference documents which makes them very significantly brighter than the screen. This creates discomfort associated with the bright objects in the user's peripheral vision. Adaptation problems are also created by viewing back and forth between the screen and reference document.

Another difficulty associated with normal office lighting is discomfort glare. Discomfort glare is created by bright lights in peripheral vision. The VDT screen is located higher in the workplace than typical desk work or typing material. Therefore, when using the VDT, bright overhead lights are considerably more in the field of view than when looking down at a desk. They can unquestionably contribute to discomfort since their brightness is usually greater than discomfort threshold levels. Similar difficulties are created by bright windows located in the field of view while viewing the VDT. Solutions to the problems include turning off some overhead light, reorienting workstations, drawing blinds or shades, or wearing visors.

Reflections in the screen of the VDT can interfere with the resolution and contrast of the material being viewed on the screen. This is also a result of poor and non-uniform lighting in the room. The most common way of resolving this problem is with the use of an anti-reflection screen placed over the VDT. Glass anti-reflection screens generally perform better at eliminating reflection and do not disrupt resolution of the screen image as much as mesh filters.

Many back and neck aches were associated with improper workstation furniture and arrangement. Chair, keyboard and screen height should all be independently adjustable so that proper posture can be maintained. Generally the chair height should be adjusted to enable a right angle at the knees, the keyboard height should then be adjusted to enable a right angle at the elbows and then the screen height should be adjusted to be 10–15° below the eyes. This necessitates adjustability in the workstation and the use of a detached keyboard. Screen heights that are too high or too low cause awkward neck positioning and result in neck and back strain.

Screen colour was not observed to play a significant role in patient symptomatology. A few patients observed after images based upon the McCullouch effect. These have always occurred with patients using green characters on a dark background. This was observed infrequently and does not represent a major difficulty.

Many of the diagnosed visual problems such as accommodative and binocular vision disorders can more easily become clinically manifest when viewing at a compromised image. Therefore, good character resolution is important for many users.

Statistical testing

Chi-square calculations were performed comparing each individual symptom (headaches, eyestrain, diplopia, blurred vision) with each individual diagnosis or treatment at the 0·05 significance level. Out of 255 chi-square calculations, less than

Table 12.5. Comparison of symptoms scores before and after treatment.

	AVG score		Paired	Significance
	pre-exam	post-exam	*t*-test value	level
Frequency	3·32	2·55	4·13	0·0002
Severity	2·67	1·90	6·23	0·0001
Total	5·99	4·45	5·71	0·0001

Table 12.6. Symptomatic relief achieved (n = 56).

Complete relief	14
Mostly relieved	20
Slightly relieved	14
No change	7
Symptoms worse	1

the amount expected by chance alone were significant when testing at the 0·05 significance level. No conclusion can be made from this analysis.

The pre-questionnaire symptoms scores were compared to post-questionnaire symptoms scores with a *t*-test. Of those who returned the post-questionnaire, there was a significant decrease in frequency and severity of symptoms, as shown by the paired *t*-test results (Table 12.5). Of the 56 respondents, 25% expressed that they attained 'complete relief' as a result of their VDT treatment and almost 36% expressed that their symptoms were 'mostly relieved' (Table 12.6). Conclusions cannot be drawn from these data, however, because of the poor response rate on the post-questionnaire (56/153).

Summary

The population of patients presented in this study is a highly selected one. They were VDT users who were experiencing symptoms and sought eye care to resolve these symptoms. This involved a great deal of self screening.

The symptoms which were observed and recorded include: eyestrain, fatigue, headaches, burning eyes, blurring of the screen during use, blurring of distance vision after VDT use, intermittent double vision, neck ache and back ache. The symptoms of discomfort can occur shortly after beginning work on a VDT or may not occur until after several hours of work at a VDT. They range in severity from barely noticeable to needing to discontinue work. The blurriness and double vision can likewise occur immediately or after several hours of work. Neck ache and back ache are quite commonly observed.

The etiology of the symptoms can be poor environmental design or an individual visual deficiency. It was difficult to assess the relative contribution of environmental and vision factors in many cases. However, it was clear that clinically diagnosable vision problems were a factor in many, if not most, patients.

The clinical diagnoses clearly indicate that presbyopia and improper correction of presbyopia are frequent causes of symptoms. There are also many younger (pre-presbyopic) patients who have inadequately functioning accommodative systems which result in discomfort. Many other visual deficiencies such as esophoria, exophoria, hyperopia and astigmatism were also responsible for discomfort.

It cannot be said whether these diagnoses are more frequent among VDT than among non-VDT users. A controlled study would be necessary to determine such. It may be that the frequency of occurrence of these clinical conditions is the same among VDT users as among the general population, but that since the VDT is such a visually demanding task, individuals with such visual deficiencies will be more prone to having symptoms if they work at a VDT. Marginal visual deficiencies are more likely to express as symptoms on a highly visually demanding task.

Acknowledgements

We thank Mark Spivac and Ken Fujimoto for their assistance in data collation and manuscript writing.

References

ANSI, 1986, *American National Standard for Human Factors Engineering of Visual Display Terminal Workstations* (Santa Monica, CA: Human Factors Society).

Howarth, P. A. and Istance, H. O., 1985, The association between visual discomfort and the use of visual display units. *Behavior and Information Technology*, **4**, 131–149.

Matula, R. A., 1981, Effects of visual display units on the eyes: a bibliography (1972–1980). *Human Factors*, **23**, 581–586.

Padmos, P., 1987, Visual fatigue with work on visual display units — the current state of knowledge. In press.

Smith, M. J., Cohen, B. G. F. and Stammerjohn, J. W., 1981, An investigation of health complaints and job stress in video display operations. *Human Factors*, **23**, 387–400.

Chapter 13

Neck and Arm Pain in Office Workers: Causes and Management

MICHAEL PATKIN

Introduction

This chapter describes a study of neck and arm pain among operators of video display units (VDUs) and others doing repetitive work in a small industrial city in South Australia and is a combination of medical and ergonomic analyses.

Apart from neck and arm pains, other common health complaints among VDU operators are eyestrain, concern about miscarriage and foetal abnormalities and facial rashes. Controversy about all of these is aggravated by several factors:

1. conflict between employers and employees,
2. lack of trust by employees, based on past hazards
3. threat to job security of new technology, and
4. lack of medical knowledge and poor results of treatment.

Emphasis has varied from one country to another, for example neck and shoulder problems in Scandinavia and Japan, concerns about eye damage and radiation risks to pregnancy in the United States, musculoskeletal conditions in Australia, and until recently little public awareness of any problems in the UK. As information is shared internationally, these distinctions are disappearing.

The Australian experience has been instructive. The terms tenosynovitis, repetition strain injury (RSI), and overuse syndrome were used at different times to refer to complaints among process workers and typists developing severe or intractable pain and weakness in the arms and hands attributed to their work.

Doctors in Australia faced an epidemic of RSI in the early 1980s which they were not trained to cope with, and their role of authority was transferred to ergonomists and to other health professionals. These in turn had temporary success by correcting obvious ergonomic deficiencies, partly through the usual placebo effect, and also because these groups opened up communication between managers and employees. Professional leadership then passed to rehabilitation services. Since then, State government programmes of rehabilitation have cost billions of dollars in their first few years. Much of this cost has gone towards treating chronic pain syndromes — conditions where pain persists in the absence of physical explanation, aching arising from normal muscle use, or exaggerated symptoms from muscle strains that would otherwise have been regarded as less significant. However, effectiveness of programmes in terms of return to work is improving, and fortunately the incidence of reported new cases also appears to be declining.

During this time various models of the 'RSI phenomenon' were developed (Wright, 1987). They included a wide range of social and psychological factors as well as medical and ergonomic ones. However, they failed to explain many of the problems of neck and arm pain, and new theories about their causes have emerged to challenge them.

The common overlooked causes

This chapter describes four aspects of neck and arm pain at work whose details have received too little attention in medical and ergonomic models. These are muscle tension, hand position, neck posture and rehabilitation.

Muscle tension

The normal amount of muscle activity needed for movement or control of posture is exceeded in several circumstances:

1. *Tense habits of work* — People have different individual patterns of muscle activity, seen in a wide range of activities such as handwriting, walking, talking, music and golf. These differences include poor timing of acceleration and speed of movement, and not exerting the right amount of force. If this force is excessive through 'trying too hard', there is more than normal muscle fatigue and ache. Often this occurs with haste, anxiety or anger.
2. *Co-contraction* — Muscles seldom contract on their own without contraction of opposing muscle groups; these correct slight errors in speed or direction through gentle braking, similar to the small adjusting movements of a steering wheel.

   'Co-contraction' is isometric contraction of active and opposing muscles (agonists and antagonists) at the same time when it is excessive. It can be demonstrated by putting the hand into an aggressive clawing position and seeing or feeling the hardness of the muscles of the forearm. Severe and continued co-contraction will cause pain which may be confused with an obvious medical condition such as carpal tunnel syndrome or tenosynovitis, especially when combined with poor wrist posture (see next section).
3. *Excess demands* — Caused by, for example, heavy levers or stiff controls, this factor is well known in ergonomic practice, but excessively stiff keys or badly maintained controls can still be found.

Hand position

Strong actions by the fingers are carried out by the powerful muscles of the forearm, which curl the fingers into a fist. More rapid, but weaker, actions are carried out by the small intrinsic muscles of the hand, which work best with the fingers bent at their base but less curled along their length. Modern electronic keyboards require a light touch, much less than for older manual machines, and therefore a straighter position of the wrist, rather than the dorsiflexed posture which favours more powerful movements by the heavier and slower muscles of the forearm used for more powerful grips and movements.

Some typists dorsiflex their wrists severely, or angle the hands outwards. To maintain these positions, the muscles which bend the wrist backwards or towards the side of the little finger, which take origin from the outer side of the elbow, must contract much more than normal.

Neck posture

Discomfort in the neck and arm is often due to a poked-forward position of the head. Such posture can be caused by glasses with too short a focal length. The mechanism of pain is probably excess tension in the nerves of the arm, like that of sciatica, when disorders of the lumbar spine cause pain in the leg.

Sudden high work load

There is a higher incidence of musculoskeletal pain, as well as accidents, when people are just back from holiday, or when the workload has suddenly increased. Even when the job design and style of work have been reviewed, experience shows that people re-starting work after an injury or accident need to begin with a much lighter schedule. This is gradually increased to normal over a period of days or weeks.

Materials and methods

The data presented in this paper consist of observations on 133 individuals, all those seen between 1982 and 1988 with work-related pain in the arm and neck in the private practice of a surgeon with a long background in ergonomics. These cases were referred as patients from other doctors, as clients of lawyers, and in a few cases as employees directly from their workplace.

There were 35 cases, or approximately one quarter, whose symptoms were directly related to work with VDT terminals, and another smaller number of clerical workers with problems from handwriting. The remaining cases covered a wide range of occupations, but with important characteristics in common.

Most patients came from Whyalla, a steel-making city of 28 000 people situated in a sparsely populated, arid area 400 km by road from Adelaide, the state capital. Some patients were referred from Port Augusta, a small city 1 h away by car. Others came from small country towns in a radius up to 600 km, or were flown from Adelaide for medico-legal assessment, or reviewed in Adelaide itself.

During the same 6 years approximately 4500 other patients were treated, so that the group under study constituted 3% of the total. A former telegraphist and a patient aged 81 who are discussed below are not counted as part of the series.

In early cases only a brief history was recorded, and much of the information desired later was not available. At the start of the study, cases of disabling pain associated with office work appeared as ergonomically interesting but not especially significant. Only as the study progressed did the significance of such cases emerge as part of a huge and expensive national epidemic.

Records were reviewed retrospectively in 1985, after 76 cases had been seen, and again in February 1988, when there had been 133 cases. A further 10 new cases seen by the time of writing are not included here. (Measurements of force in this study are recorded in Newtons; 1 N is approximately equal to 100 g weight, or just over 3 oz. An American 5-cent piece, or 'nickel', and an Australian 10-cent piece each weigh about 5·5 g.)

Early observations

Observations on the first 75 cases were presented to a state meeting of the Australian Physiotherapy Association in 1985, under the title of *A Muddled Label*. Each case was summarized in 40 words or less. Key expressions were then used to classify them into 17 descriptive groupings or clusters (Table 13.1).

The nine cases in the subgroup of distorted symptoms (12%) are listed individually (Table 13.2). They were cases where there was no connection evident between the stated cause and the long-standing pain. They were important to identify as early as possible, because they were difficult or impossible to help medically, and they gave other cases a bad name.

Table 13.1. Descriptive groupings, first 75 cases.

Group	*No.*
Chronic pain syndrome, permanent	6
Chronic pain, for further evaluation	10
Chronic pain, recovery	1
Chronic pain, cure by job change	3
Chronic pain, ergonomic cure	2
Epicondylitis, mostly cured medically	5
Retired for other reasons	1
Misrepresented symptoms	9
Doubtfully genuine	4
Temporary complaint	5
Excess or faulty typing/writing	10
Musicians	4
Workplace with much manual handling	3
Data-entry workplace	2
Non-work problem	3
Not clear	6
Complicated other problem	1

Table 13.2. Distorted symptoms.

No.	Stated cause, symptoms
1	Polishing jewelry 1h weekly
2	Blood test followed by 'wrist paralysis'
3	Supermarket work causing breast pain
4	Using microfiche cards intermittently
5	Arm pain from driving on bad roads
6	Hammer strain, glove numbness, refused rehabilitation
7	No cause, multiple pains, childhood traumas
8	Elbow bump, inconstant signs
9	Arm pain, recent cot death of baby

Three cases labelled 'ergonomic success' are described in more detail (Table 13.3). Each of these had chronic pain cured by ergonomic analysis and alteration of their activity. Normally two of them would otherwise have undergone surgery in an attempt at cure.

Table 13.3. Ergonomic cures.

Cause	Mechanism	Cure
Writing too hard	Habit, stale carbon paper	Fountain pen
Poor supervision	Bent neck	New glasses, desk
Kneading dough	Bent wrists, age 81	New hobby

Case 1. Writing too hard

The 40-year-old female manager of a fashion boutique had a classical De Quervain's tenosynovitis of the right wrist, with symptoms for 2 years despite physiotherapy and anti-inflammatory drugs. When she was referred for surgery, there were typical swelling and crepitus which left no doubt about the diagnosis. However examination of her handwriting style showed that she pressed very hard with a pen. She was prescribed a fountain pen, lost all her symptoms in 5 days, and had no need of surgery.

Follow-up examination at her shop showed she wrote dockets for 1·5 h per day using stale carbon paper. When writing she pressed with a force of 6 N. This was measured on a Chatillon push–pull gauge mounted immediately next to her desk, with her hand supported in the usual way. After she started using a fountain pen, the force reduced to 1·5 N. This was still excessive, as the minimum force required to write with a fountain pen is less than 0·01 N (under 1 g weight). There was a family joke that her husband was able to hear her writing across the other side of their thickly carpeted living room.

Case 2. Failure of supervision

A 50-year-old female clerk at my hospital was off work for 6 months because of neck pain. Ten years earlier she had a cervical laminectomy and fusion for spondylosis. In the previous year she had been seen by over a dozen doctors, including two neurosurgeons, and two orthopaedic surgeons on behalf of an insurance company. Though her general health was otherwise good, she was unable to do her housework and had no social life.

Her job consisted of copying numbers from a book of medical benefits fees to accounting forms. An ergonomic examination of her workplace showed she was hunching forward to work, and the office arrangements had at least seven faults:

1. desk too low,
2. chair not adjustable,
3. source documents flat on table,
4. writing surface flat instead of sloping,
5. inadequate lighting
6. reading glasses focused too close, and
7. lack of supervision of her working posture.

A report of these faults was ignored by management for several months, and she was off work most of this time on compensation payments. In the meantime her husband, a technician, made an adjustable board to support the documents she was reading and writing on, and her desk and chair were replaced with adjustable ones. She returned to normal work after this.

Follow-up 1 year later showed she was now quite free of symptoms, and her personal and social life had improved remarkably. Sad to relate, 2 years after successful rehabilitation her job was computerized with no ergonomic supervision of the changeover. Within a few days her old neck problem had recurred because of poor workstation design and bad neck posture. She was off work for 2 months because of further symptoms and then retired permanently on an invalid pension.

Case 3. Not kneading the dough

An 81-year-old female, mentioned earlier, was referred with clear-cut symptoms of bilateral carpal tunnel syndrome for 1 year with typical distribution of night pain. On taking a detailed history, it emerged that she had taken up home bread-making as a hobby shortly before her symptoms began. When she mimed the movements of kneading the dough, it became clear she dorsiflexed her wrists forcibly. After she gave up her hobby, with regret, she was pleased to find her symptoms disappeared within a few days.

The present series

With increasing experience with the first 76 cases, a more detailed history and examination were recorded for later patients. Towards the end of the study, a data base was created on a personal computer using dBASE3, with 26 fields for each case record. Six fields had clinically less relevant information (number, name, date, doctors

Table 13.4. Clinical information.

3. Age	15. Hypoaesthesia
6. Job title	16. Brachial plexus test
7. Likely ergonomic factor	17. Co-contraction
8. Past history	18. Psychological factors
9. Symptoms	19. Diagnoses
10. Complaint of swelling	20. Management
11. Duration	23. Outcome
12. Stated disabilities	24. Comment
13. Exaggerated disability	25. Organic or functional
14. Signs (general)	26. Notes

and solicitors involved, and employer) leaving 20 fields with clinically relevant information (Table 13.4).

At first this tabulation was a clumsy tool for analysing data because the study was retrospective. The history had not included many of these items, and criteria for them had not been defined at the start. A study of individual factors sometimes required creation of an additional field, manual transfer of data from other fields, and review of all the records in the series one by one. This was done several times. These categories will be revised for future studies.

In some other respects the direct manipulation of the data base became a particularly useful tool. A standard terminology evolved by repeatedly browsing through records in an alphabetical order for the factor under review and altering inconsistent terminology, and some information was extracted by highlighting particular records in the printout.

Observations

There were 30 cases seen only once. Reasons for this were that their complaint was minor and easy to resolve, they had been referred only for medico-legal assessment, the patient chose not to return for further treatment, or they had been advised management elsewhere.

There were 103 cases reviewed on more than one occasion, and many of these were seen between 5 and 10 times to monitor their course and management.

Table 13.5. Occupational groupings.

Occupation	No.
Office workers (clerks and journalists, typing, 35; writing, 10)	45
Factory workers	38
Service workers (food, 11; cleaning, 10; sales, 9)	30
Switchboard operators	5
Music teachers, students	4
Miscellaneous	11

Occupational groupings

The main occupational groups were office workers (45), factory workers (38), and service workers in food, cleaning and sales (30). Others were switchboard operators, music teachers or students, and miscellaneous (Table 13.5).

Muscle tension

These results were reported briefly at the recent International Ergonomic Congress in Sydney (Patkin, 1988a).

The cases in whom excess muscle tension was detected were office workers, cleaners and a labourer. While the main stated cause for pain among office workers was keyboard work in 35 cases, there was also a group of 10 cases who considered the cause of their problem was handwriting at work.

Tense habits of work were assessed on criteria which were added to as the study proceeded. These now comprise general habitus, handwriting faults, and other movements at work.

General habitus

The arm movement of subjects was observed as they walked into the examination room. They were asked to swing their arm loosely and then observed for free movement of the elbow and wrist. The subject's forearm was supported loosely, and then allowed to fall under its own weight. Subjects were considered tense if the arm stayed momentarily in position before falling.

Handwriting faults

1. *Pressing too hard with the pen* — This was assessed by having the subject write on ordinary paper rested on a folded handkerchief. If the subject pressed too hard, the paper was indented, and this was evident on looking at and feeling the other side of the paper. The greater the force, the more severely was the paper indented, and in some cases it was actually torn.
2. *Gripping the pen too hard* — The index finger-tip was bent back passively and the skin blanched over the back of the interphalangeal joint. Figure 13.1 contrasts the index finger posture of a forceful pen grip [Figure 13.1(a)] with the superior posture of a relaxed grip [Figure 13.1(b)].
3. *Co-contraction* — This was manifest when the extensor muscles of the forearm were contracting too firmly. This could be felt, and in thinner subjects could also be seen.

   A pocket electromyograph (EMG) made by Group Occupational Health Services of Sydney was also used to demonstrate this, and made it much easier to explain to the subject what the problem was. This device was attached to three surface electrodes on the subject's forearm, of which two were a few centimetres apart over the line of the extensor muscles under test, and the third

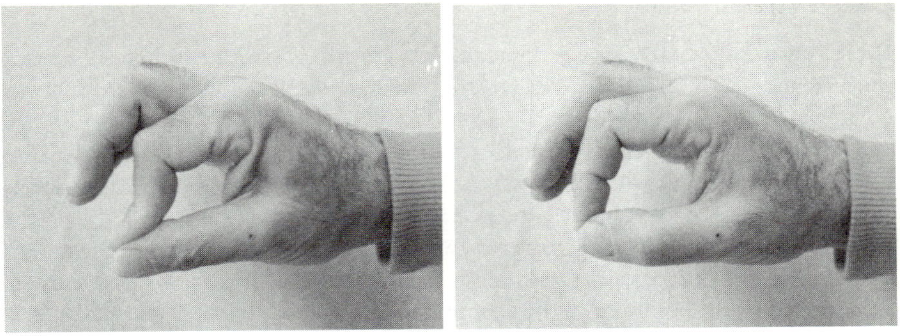

a b

Figure 13.1. (a) Forceful and (b) relaxed pen grip postures (reproduced with permission from Modern Medicine of Australia).

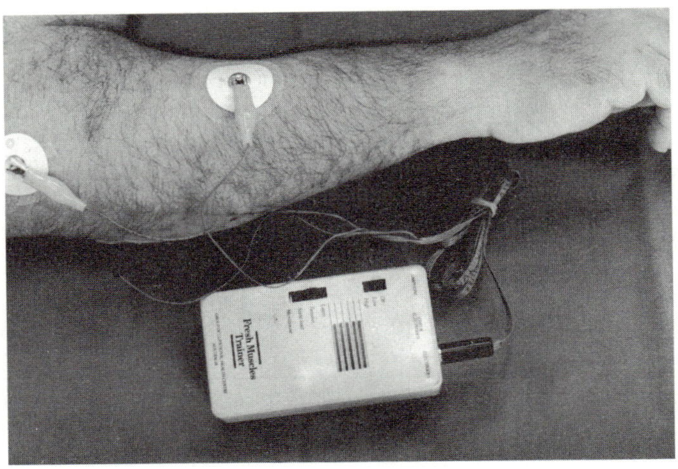

Figure 13.2. A popular Australian, pocket-sized, EMG and electrode siting.

was an indifferent electrode a little away from this line (Figure 13.2). Ingenious and compact circuit design resulted in bleeping when muscle tissue contracted, and the pitch and frequency of the bleeps increased as the muscle activity increased. The small size, freedom from extraneous signals, and reliability of the EMG made it an impressive and effective tool both for diagnosis and understanding by the subjects.

Other movements at work

Subjects were asked to mime their movements at work, or observed while duplicating them. A judgment was made as to whether the movement was relaxed, tense or not obviously either.

Several of the typists were observed operating a keyboard. A note was made of how hard they pressed the keys by listening to the loudness of the impact, as well as noting their degree of wrist flexion or extension.

Those subjects who were considered to have tense patterns of muscle movement were shown the difference between tense and relaxed handgrips in writing, how to press the pen with less force, and a demonstration of co-contraction with the EMG. Some typists were given a trial of EMG to show them the effect of varying wrist posture on muscle activity.

Office workers who wrote a lot and had tense writing habits were told which brands of pen wrote most easily. A separate study was carried out of 23 different models of ball-point pen writing on paper supported on cloth, as described earlier. Measurements were made of how much force was needed to write legibly (0·03–0·25 N), to indent paper (over 0·2 N), to tear it (over 0·6 N), and to make legible carbon copies (variable). Fountain pens write with under 0·01 N. Some brands of ball-point pens now available write consistently with a force of less than 0·04 N.

Results of treatment of office workers with excess tension

There were 37 cases of office workers using excessive force when typing or writing, according to the criteria above. Only 22 cases out of the 37 were followed up adequately. Most of these improved with simple advice (Table 13.6).

Table 13.6. Results of treating excess muscle tension (including dorsiflexion) in office workers.

Activity	Number	Followed up	Improved	Not improved
Keyboard	27	13	7	6
Writing	10	9	8	1

TYPING

Among the 35 keyboard workers in this study, eight had neck pain as the leading symptom, and 27 complained mainly of arm or hand pain. Among the latter group, 14 were seen only once, leaving 13 cases with hand and arm pain treated and followed up. Of these 13 cases, seven were cured by simple advice to buy better-quality ball-point pens and to grip and press with less force. It is likely that several of the cases seen only once also recovered. Of the six cases which failed to improve with simple advice, four were involved in protracted litigation or had disputes at work, or had significant psychological problems, and the other two had significant psychological problems in their personal life.

These results include cases where the excess muscle tension was due to dorsiflexion of the wrist, described in the next section.

WRITING

All 10 clerks with pain from handwriting had a tense technique. Of this group, all except two lost their symptoms in 1 or 2 weeks. One was not followed up, and the other was angry with his employer over past events.

Results of treating non-office workers with excess tension

Isolated cases of muscle pain associated with excess strain were seen in four other activities — cleaning with electric floor-polishers, hammering, morse-code operation, and music — as well as the amateur breadmaker described earlier. The numbers are too small to have significance, but analysis of their activities appears to support the general thesis in this section.

CLEANING

There were two cases of lateral epicondylitis in this group. A work-site inspection showed that both had occurred in cleaners who pulled a heavy commercial polishing machine up a step because there was no sloping ramp for them to use.

Their supervisor, who had no problems of arm strain, had a technique of pulling the machine up the step by keeping the elbow straight and balancing the machine with her body weight (Figure 13.3). This allowed the extensor muscles of the forearm to stay relaxed.

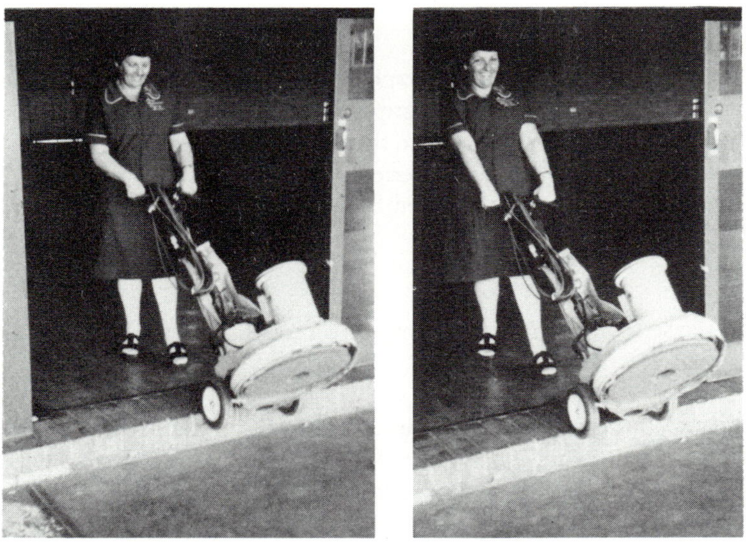

Figure 13.3. Two ways of pulling a heavy polisher up steps.

Simple observation of the arm action required shows that if it was carried out with the elbow slightly bent, it requires strong contraction of the forearm extensors attached to the lateral epicondyle. It is easy for the seated reader to confirm this phenomenon by pulling on the edge of the seat, while palpating the extensor muscles of the forearm with the other hand, first with the elbow straight and relaxed, then with it bent.

There is a second risk factor related to skill for cleaners using floor polishers. Normally the rotating head of the polisher is kept nearly flat on the floor. Tipping the polisher gently towards and away from the operator makes it move to the left and right in turn, giving arcs of movement across the floor. Tipping it slightly to one side and then the other at the end of each arc gives a more graceful and easier figure-of-eight movement. If the operator tips the machine too much, it means she has to 'fight' the polisher and grip it more fiercely.

Both of the cases in this series were involved in litigation, and in neither case did the symptoms improve over many months. A third cleaner, who also failed to improve over several months, had the habit of holding the wrist flexed all the time. If she held a shopping bag, it took extra effort to keep it out of the line of gravity. She held the wrist flexed severely, whether holding a broom or a shopping bag.

Hammering

There was one case of a male ex-clerk aged 36, who developed pain in both forearms while cleaning tough grease from inside a large furnace with a chipping hammer. When asked to mime the technique he used, he pushed the hammer by extending the elbow with a short stroke and a stiff arm, instead of swinging in a loose movement from the shoulder joint.

Morse-key operators

At a local historic demonstration of telegraphy, two former operators showed distinct styles of hand-grip. The operator with a grip tense enough to hyper-extend the tips of the index and middle finger-tip [similar to the posture shown in Figure 13.1(a)] used to suffer with forearm pain at work, referred to as 'glass forearm' or 'telegraphists' cramp'. The operator with gently curled fingers [similar to the posture shown in Figure 13.1(b)] had no such pain.

Musicians

All four in the series had unusually strong co-contraction when their handwriting style was tested.

To summarize this group, in this study, 15 cases who were followed up out of a series of 133 cases with work-related pain, or 11%, gained relief by learning to relax their habits of work. The incomplete follow-up means this success rate has been underestimated, but many cases also demonstrated tension of nerves in the neck and arm (see next section), so no conclusions can be drawn about the significance of muscle tension on its own.

Hand position

As mentioned earlier, subjects had been asked to mime the action of typing, and then observed while actually typing. Wrist posture was considered faulty if there was dorsiflexion of the wrist beyond the neutral position, or if there was more than minimal ulnar deviation of the wrist.

Although observed on only a few occasions, every case in which the wrist was dorsiflexed when typing showed increased muscle tension compared with that when the wrist was straightened. This was obvious on palpating the extensor muscles of the wrist, either by the examiner or the subject's other hand, and in a few cases on using the pocket EMG. The outcome in these cases has been incorporated in the data for 'excess tension' in the previous section.

Similar excess muscle tension could be demonstrated if the wrist was deviated to the ulnar (little finger) side, either because the keyboard was to one side (e.g. to make room for a document holder), or habit, or because the subject had a wide trunk, so that the forearm was at an acute angle to the keyboard instead of a right angle.

One related phenomenon was seen at an office among a group of keyboard workers at a large remote worksite, whose data were not included in this series. The accounting system used had recently been converted from a manual system to a computer-based one. Of 50 operators, 11 had begun to complain of forearm pains similar to those described earlier within a few weeks of starting to operate keyboards.

In this system, all alphabetical fields were entered in upper case. To save using the numeric pad, the shift lock key could be operated to reassign ten of the alpha keys temporarily for numeric entry, and operated a second time to enter the numeric field. This key was at the extreme right end of the keyboard and operated by the little finger. At the time of the site inspection only 2 out of 50 operators in the area were entering data. However two quite different techniques of operating the key could be seen from as far as 5 m away. Using the 'forearm shift' technique, the operator moved the whole forearm to the side, like a pianist playing, to position the little finger over the shift lock key. In the 'little finger stretch' technique, the forearm moved less or hardly at all, and instead the little finger was stretched sideways.

It appeared likely that the second style was the one used by those complaining of pain. Testing muscle tension by palpation and by EMG confirmed the excess muscle tension on the outside of the forearm with the 'little finger stretch' technique.

Prior to the inspection and subsequent report with recommendations for supervision of keyboard technique as an initial step, the employees' union had pressed for redesign and supply of modified keyboards, with the return key placed next to the alpha keys. The supplier had quoted a cost of A$100 000 for this modification.

Observations and discussion with operators at the time suggested the alteration in typing technique could be effective in solving the problem. Follow-up on this group was not available, but its success seemed likely on the basis that the operators could understand the cause of their new pain and what they could do to prevent it.

Neck posture

It was only in the latter part of the series that a test was included for brachial plexus

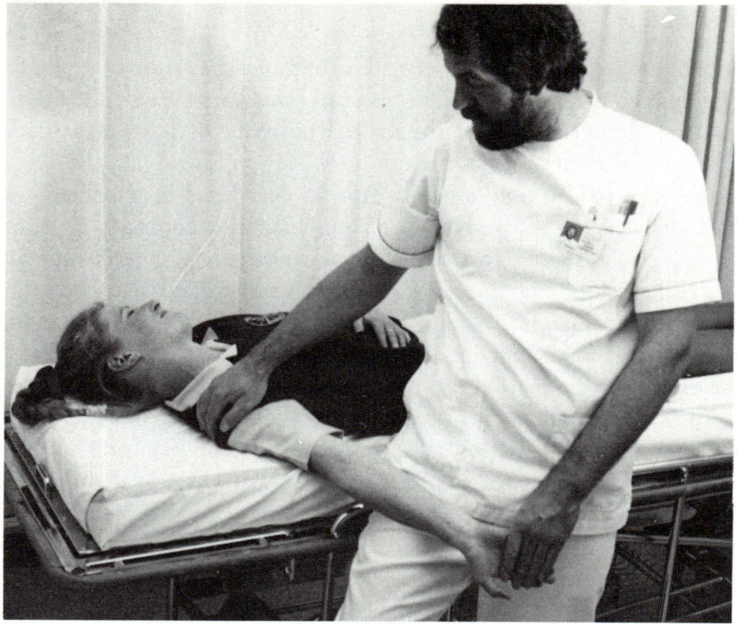

Figure 13.4. Brachial plexus tension test — one of several steps (reproduced with permission from Modern Medicine of Australia).

tension (BPTT). This test was devised by Elvey (1979), a physiotherapist in Western Australia, in 1979. In this test the large nerves of the brachial plexus are stretched by putting the arm in a suitable position. This is similar to stretching of the sciatic nerve in the well-known straight-leg raising test, a common clinical test for lumbar spinal causes of pain in the leg.

The version of the BPTT used here was simplified from the version devised by Elvey (Figure 13.4). The steps are as follows:

1. The subject lies flat on the back on an examination couch.
2. The head and neck are bent (not turned) to the side away from the examiner.
3. With the subject's shoulder out to a right angle, the examiner holds the subject's forearm with the palm outwards (supinated), straightens the elbow, and steadies the upper arm against the examiner's thigh.
4. The examiner pushes the patient's shoulder down towards the feet.

Each of these steps applies tension to the nerves of the arm along their whole length, a fact which becomes obvious if these manoeuvres are carried out when these nerves are exposed at operation.

The BPTT is positive if it reproduces the pain and tingling in the site usually noticed by the patient in relation to work. These symptoms should disappear when the patient's shoulder is allowed to elevate to its relaxed position. The commonest site where symptoms are felt is over the 'extensor pad'. (This refers to the prominent muscular area a few centimetres below the outside of the elbow.) In some patients these symptoms occur at the front or back of the wrist.

This test may be positive in one or both arms. There is also an incidence of false positive or 'physiological' results. In normal people it is often possible to produce similar tingling and discomfort in some part of the arm, but when the site is the same as that of work-related symptoms the pains are likely to respond to the simple treatment measures outlined later in this chapter.

Elvey now distinguishes between different parts of the brachial plexus and different nerve trunks by flexing the elbow (which relaxes the median nerve but tenses the ulnar nerve), adjusting neck position, and other manoeuvres which were not included in this series (R. L. Elvey, personal communication).

Findings using the BPTT

This test was only carried out in 54 cases in the latter part of the series after it had become known to the author. It was also omitted in a few later cases where it was clearly not relevant, or where the subject happened to be seen at the worksite and suitable examination facilities were not available (Table 13.7).

Table 13.7. Findings using the BPTT.

Total cases		133
Positive	31	
Negative	9	
Non-organic	1	
Physiological[a]	13	
Total cases tested		54

[a]Physiological, production of aching or tingling in a different area to the one in which the symptoms had been reported.

The management of those cases with neck pain was to demonstrate the relationship of their arm symptoms to their neck posture so that they were able to understand it clearly, and then to advise them on neck posture, neck exercises, and where appropriate the need for glasses of longer focal length. Some subjects were referred for manipulative treatment.

Of the 31 subjects in whom the test was positive, 16 were known not to have gained relief during the period of the study. There were eight cases where the outcome was favourable, and seven in whom the outcome was not known. The favourable results occurred where subjects followed advice to avoid slumping their neck and to carry out neck retraction exercises (McKenzie, 1983) as discussed below.

Of the nine subjects in whom the BPTT was negative, one was not followed up, four improved and four did not.

Many of the cases also had evidence of excess muscle tension, as listed earlier. The overlap between these two factors means it is not possible to assign significance to them individually.

Rehabilitation

Simple cases had no formal rehabilitation. In many other cases, an attempt was made to contact the employer with a view to correcting ergonomic problems with the job and allowing the employee to return to modified work. Often this meant starting with 2 h per day and increasing to a full day's work over several weeks.

In later cases, job design was assessed by ergonomists from an outside rehabilitation service, to which patients were referred if they were not making good progress.

There were 30 cases seen only once. Of the remaining 103 cases, 40 appeared to achieve complete relief, 50 were unchanged or worse, five improved partly, and the outcome was not clear or unknown in eight cases (Table 13.8).

As expected, the longer the duration of symptoms, the more likely it was that resolution would not occur. The transition between the two groups was generally at 2 months, although there were exceptions. Cases with symptoms for 1 month or less recovered, with one exception. However there were 11 cases with long-term disability after 2 months of symptoms, and nine cases with symptoms for 1 year or longer who returned to full-time work, sometimes their original job. Two of these cases were described earlier as 'ergonomic cures', and one with symptoms for 5 years was cured by a carpal tunnel operation to decompress the median nerve.

Table 13.8. Duration of symptoms when first seen versus outcome.

Duration	Relief	Partial	None	Unknown	Total
<1 m	9	1	2	—	12
1–2 m	12	1	9	2	24
3–12 m	10	3	19	3	35
>12 m	9	—	20	3	32
Total	40	5	50	8	103

Discussion

Much of the motivation for this study arose from the author's frustration in attempting to diagnose and treat increasing numbers of patients referred with work-related symptoms which did not fit into current medical models. Some of the interest in these problems developed from earlier observations on the nature of skilled hand function. A further stimulus to work in this area was the serious and growing public controversy and the escalating costs in working time and diversion of resources to deal with it.

The Australian experience

Complaints of pain among VDU operators and others doing repetitive work have been costly in both human and financial terms. In Australia the cost to the economy, including direct working time, compensation to employees, advisory services, and diversion of management resources, could be measured in billions of dollars.

The term repetition strain injury or RSI was applied widely in Australia to occupational pains in office workers and others after its first appearance in a government report (NHMRC, 1982). Similar terms used earlier were 'repetition injury' (Ferguson, 1971) and muscle repetition injury (Taylor *et al.*, 1981). Its use became widespread in the next few years. Its history as a social phenomenon has been reviewed in detail (Spillane and Deves, 1987), as well as its recent decline in a large national organization (Hocking, 1989),

Australian awareness of occupational pains and strains, together with other health and safety issues at work, came about largely due to Dr John Mathews, a graduate in political science. He became the first director of the Occupational Health and Safety Unit of the Australian Council of Trade Unions in 1981. This unit published frequent bulletins on health and safety issues at work (*Health and Safety Bulletin*, periodical) which had wide circulation and great influence. Mathews was a competent analyst, publicist and writer (Mathews, 1985), but his political commitment at times appeared to outrun the scientific data.

As a result of the influence of Mathews and others, tenosynovitis was frequently diagnosed in factory process workers and typists. Young women wearing forearm splints became a common sight in medical and paramedical waiting rooms, in newspaper photographs and occasionally in city streets. It became clear that this diagnosis was wrong in nearly all cases. The epidemic of reported RSI cases in Australia was a regular feature in the national media and then appeared to become less fashionable.

A vigorous debate developed between the proponents of two views 'it's all in the head' versus 'it's all in the body'. Doctors became caught up in the same bitter debate, with diagnoses such as conversion hysteria on one side and 'overuse injury of muscles' on the other. Numerous theories appeared to explain the epidemic, including fluoridation, neck problems, muscle tension, malingering and psychogenic pain.

Each of these views had supporters who excluded all other viewpoints. They ignored emerging factors such as congenital ligamentous laxity, carpal bone instability, previous injury to the neck or fracture of the forearm (even in childhood), and co-incidental specific problems such as capsulitis of the shoulder and lateral epicondylitis, either arising spontaneously or from identifiable causes.

The diagnosis of common conditions was sometimes missed by doctors irritated by increasing numbers of patients, often self-referred with a diagnosis of RSI, so that simple tests for definable injuries such as lateral epicondylitis — lateral elbow pain on resisted extension of one or more of the central digits (Figure 13.5) — were not carried out. A fresh clinical approach to individual patients was needed (Patkin, 1988b).

It gradually became clearer that RSI, overuse syndrome and similar expressions were unsatisfactory. They were confusing and damaging umbrella terms which refer to a wide range of different conditions. Their common features were pain related or attributed to work, but like other medical and social entities there is a central group of shared features whose different patterns make rigid definition a problem.

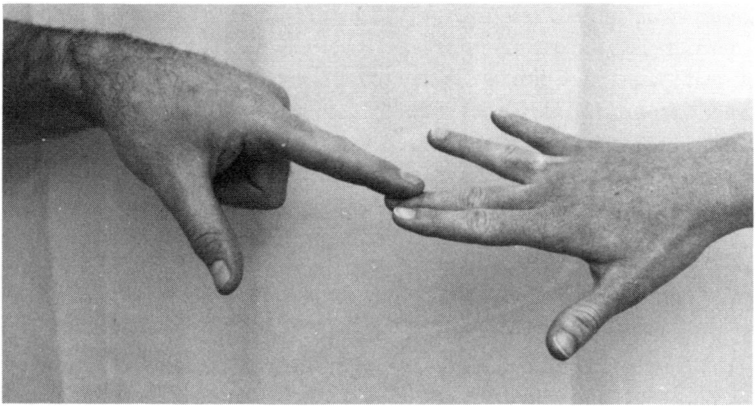

Figure 13.5. Simple test for lateral epicondylitis (reproduced with permission from Modern Medicine of Australia).

The very choice of terminology became a prejudgment, an expression of viewpoint or prejudice. If a person with professional authority uses the term 'injury' in this context, this is likely to be understood as implying a physical disruption or displacement (by contrast with 'injured feelings'). If a neutral term like 'complaint' or 'pain' is substituted, there is more likely to be the belief or understanding that there is some temporary and quickly reversible condition.

Against this confusing background, the purpose of this current report is to draw attention to specific physical factors which have been neglected in the past. Great attention has been given to details of seating, screen characteristics and other ergonomic factors, but the factors considered here seem to have fallen into a gap between medicine and ergonomics.

Regrettably this retrospective study does not distinguish between the roles of the different factors analysed as clearly as was hoped initially.

Muscle tension

It is an everyday observation that individuals vary in the pattern of their muscle movements, e.g. in walking, handwriting, speech, sport, music and work. Poor skill at work has received scanty recognition in theory or practice. A case of tenosynovitis in a shoe assembler, probably due to excessive force, was mentioned by Welch (1973). Nine cases of work pain from tense habits associated with poor skill were recorded under the label of 'trying too hard' (Patkin, 1984).

Earlier observers of cramp in writers and telegraphists considered it was due to involuntary muscular contraction (Gowers, 1888). However subjects in this series were able to learn a more relaxed writing style using the techniques described. In sport and music it is common to re-train skilled movements to be more relaxed and efficient.

This ability to relax muscles not needed for the action is seen in videotapes of competitive downhill skiers, when the hamstring muscles (at the back of the thigh) are not needed to maintain knee stability and can be seen to relax, and in the triceps muscle of a cellist who is making a confident 'ballistic' bowing movement.

Timing

Another aspect of strain and excess muscle tension is 'timing', a term which does not appear to have been defined in the ergonomic literature. Good timing is the rate of application of force which maximizes the transfer of energy. An example is pushing on a child's swing when it is moving away from the pusher. A second example is using a hammer, accelerating its movement steadily instead of with a jerk at the start, which appears to have been the mechanism in the 'hammering' case described earlier.

Timing also includes the sequence in which different muscle fibres contribute to an action. This is seen in competitive weight-lifting, when the muscles of the back come into play with their slower initial acceleration before those of the quadriceps in front of the thigh.

An area of great interest for testing this concept is doubtless that of manual handling and the prevention of back injuries, when taken in conjunction with other factors already recognized.

Further aspects — EMG, equipment, maintenance, anger

Undoubtedly the development of portable inexpensive EMG machines has been a major factor in clarifying the role of excess strain, and should prove invaluable in retraining keyboard operators and workers in other areas.

Some of the blame for bad skills and bad writing habits can be put on manufacturers of ball-point pens of poor quality which need more force for the ink to flow. Even if the ink flow fails only occasionally, writers get into the habit of pressing harder all the time in order to try and forestall this. Another factor may be the changing activities of schoolchildren. Poor motor skills, including handwriting, may be partly the result of spending their time watching television and playing with toys operated by punching simple controls, instead of hobbies requiring delicate hand activity, and playground games which develop fine motor co-ordination and timing (Patkin, 1987).

Skill includes much more than efficient patterns of muscle movement. It is also how people choose their posture at work within the range of constraints, what equipment they choose, and how they maintain it. 'The time-served craftsman has the ability to accomplish a whole repertory of tasks, and to determine the means for achieving them.' (Seymour, 1966).

In factory work the choice of equipment and how well it is maintained is outside the control of the operator. However while stiff controls on machines are well known to ergonomists as a cause of muscle load and strain, they are rarely a problem with keyboards. The ISO recommendation (ISO, 1971) of a key stiffness in the range of 0·25–1·25 N is clearly too broad, and most keys require a force of 0·6–0·9 N to operate, in line with recent studies (Greig and Caple, 1986). Nearly all commercially available keyboards now fulfil this requirement, but some numeric keypads used for rapid data entry still require a force well over 2 N to operate.

Such levels of force exerted by the fingers are surprisingly not a problem in the apparently delicate work of pressing on the strings of a violin, where they have been estimated to be in the range of 0·5–1·5 N (Basler and Kawaletz,1943).

One factor noted but not analysed in this series is anger. It is an everyday observation that some angry people become physically tense, for example in their body posture, when speaking, writing, pressing an elevator button, or operating controls. If anger causes pain (rather than the other way around), it will obviously be important to manage as a primary factor.

Hand position

The traditional wrist posture for typists, dorsiflexion, was discussed in the introductory section. It was analogous to the power grip of the hand, which is much weaker if the wrist is straight or flexed. Like bad neck posture discussed immediately below, bad wrist posture may be due to bad training or habit, or poor design of the workstation. However, it has more in common with the tense habits of work discussed above. Both are examples of pain from excessive muscle activity rather than pain from tension on arm nerves.

Favourable data are accumulating following several ventures to redesign the traditional keyboard to avoid ulnar deviation of the wrist (Malt, 1977; Grandjean *et al.*, 1985; Rose, 1986). It is probably too early for their wider acceptance. In a new development which extends the concept of wrist rests for typists, Stack (1988a, b) has argued strongly for slanting the keyboard away from the operator, which also flattens the wrist and the fingers. She asserts that this has been a major factor in solving the problems of RSI in the public service in Tasmania.

Although the importance of not flexing the wrist for rapid runs on the piano is well known to pianists, such skill factors are not taken into account by observers who claim there are microscopically visible changes in tissues in overuse syndrome (Dennett and Fry, 1988) and who rely on a controversial prescription of 'radical rest' for 'healing'.

While many keyboard characteristics have been studied in the past (Cakir *et al.*, 1980) one not yet analysed is the effect of lack of resilience of the electronic keys. This imposes a sudden stop to the movement of the fingers which is transmitted to the forearm if the fingers are curled rather than straight.

Neck posture

Pain caused by neck problems is common.

> At any given time, as many as 9 percent of the adult male and 12 percent of the adult female population is experiencing some discomfort in the neck with or without associated arm pain, and 35 percent of us can recall having such an episode. (Hult, 1954)

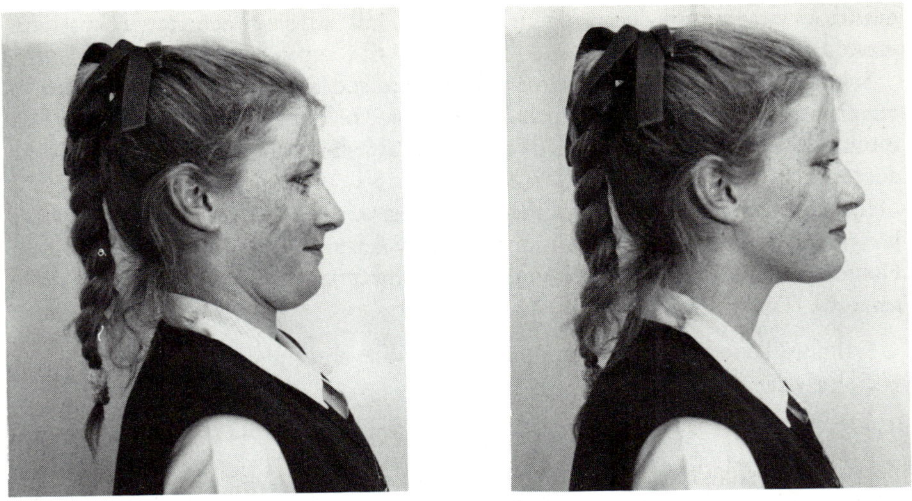

a b

Figure 13.6. Neck exercise: (a) retraction, (b) protraction, to relieve brachial plexus tension (reproduced with permission from Modern Medicine of Australia).

Understanding of neck problems is poor, and they are not taught well to medical students. Advice is available in conventional medical textbooks, from doctors taking a non-orthopaedic approach such as Cyriax, rheumatologists such as Hadler, and many others.

The approach in this series has been to look for the simple remedial factors described earlier. The exercises advised for neck problems included especially retraction [pulling the neck backwards while keeping the head erect Figure 13.6(a)] and protraction [Figure 13.6(b)] for several seconds every few minutes or hours during the day. Patients were advised to buy a copy of a simple self-help book (McKenzie, 1983).

Other advice was to sleep with a low pillow, and to attempt to maintain a neutral neck posture. If simple measures did not solve the problem, then appropriate further help was sought, from professionals such as a rheumatologist, chiropractor, masseur or psychiatrist.

Rehabilitation

Formal rehabilitation services were set up in Australia approximately 20 years ago. They were largely organized along traditional medical lines rather than common vocational ones. With the emergence of the RSI epidemic, these services became adapted to treat people with problems of work-related pain and strain.

Since the early 1980s, after government intervention, rehabilitation has become a multi-million dollar business. In some cases large business organizations and government authorities set up their own internal organizations, perhaps using some outside services as well, such as specialized ergonomic advice or contract occupational medical facilities.

Some self-styled ergonomics experts appeared, and a large industry developed to manufacture ergonomic furniture, document holders, wrist rests, anti-RSI spectacles, and other devices. Others who prospered included some lawyers and doctors.

In the last year in Australia there has been a severe shake-out among these professionals, with survival of consultants partly dependent on acceptance by unions. However, important and increasingly successful principles of rehabilitation have emerged. These include:

1. information and appropriate training at all levels of management,
2. early reporting of cases,
3. careful and detailed assessment of affected employees, typically taking several hours,
4. co-ordination of medical, paramedical, management, social, and other resources,
5. ergonomic job analysis,
6. modifying work to match the capacity of the rehabilitee,
7. monitoring individual progress, and setting realistic goals, typically with a return to full work over several weeks,
8. monitoring overall Occupational Health and Safety (OH&S) statistics carefully, and
9. accepting OH&S as a responsibility of line management, and the relationship between accident rates and good industrial relations and management techniques.

Managers of large organizations such as factories and offices had been frustrated by their inability to cope with the epidemic of reported cases. Many of these were written off with compensation pay-outs. Some large firms developed rehabilitation services with token jobs. When these failed to achieve return of affected employees to work, more skillful and professional approaches to job modification and graduated programmes of increasing workload were developed.

Follow-up and statistics for success rates of rehabilitation in this series are not available, but there is an impression that current rehabilitation methods have proved successful.

One successful form of intervention has been to employ an occupational physiotherapist to monitor staff before they made formal complaints of symptoms, and to detect and correct potential problems, such as bad seating, job design, work habits or neck posture.

Other factors

Psychogenic pain

The diagnosis of psychogenic pain is notoriously difficult. It is made partly by exclusion, and partly on positive evidence (Engel, 1959). Where organic and psychological causes for symptoms co-exist, the problem is of course much harder. While observations were made in the present series, results have not been analysed for the present study except to note the relation of anger to muscle tension. However, there often appeared a progression from acute physically-based symptoms, anger, or both, to chronic pain without evident physical basis. There were other cases which had no physical basis from the start, as shown by the group of cases with distorted symptoms.

Dogma and social attitudes

Professionals, politicians, journalists and many others in Australia made numerous public statements about RSI during the years of this study. Role models for pain behaviour are well known in other situations, and unions failed to understand that reversible short-term aches could develop into chronic pain problems without a physical basis if victims were repeatedly told they were injured. These were probably fuelled further by suspect assertions of proof of non-specific overuse injury, based on electron microscope study of tissues from muscles which have been rested for many weeks (Dennett and Fry, 1988) and mirrored by publicity for controversial conditions of vague pains attributed to virus infections such as benign myalgic encephalomyelitis and 'chronic fatigue syndrome'. There was also a delay in appreciating the role of poor management style in generating these problems. Such factors are outside the scope of the present discussion.

Conclusions

Neck and arm pain are common symptoms among VDT operators. In this study a combined ergonomic and medical approach suggests that common but preventable causes include bad habits of work, hand position, neck posture and poorly planned rehabilitation. The data in this study provide some support for the role of these factors, but the numbers of cases are small, and more carefully planned prospective studies of these factors are needed quickly.

References

Basler, A. and Kawaletz, K., 1943, Uber den Einfluss des Konstitutionstypus auf die Technik des Geigenspiels. Arbeitsphysiologie 12. Quoted in: *The Physiology of Violin Playing*, O. Szende and M. Nemessuri, 1971 (London: Collett's).

Cakir, A., Hart, D.J. and Stewart, T.F.H., 1980, *Visual Display Terminals: A Manual Covering Ergonomics, Workplace Design, Health and Safety, Task Organization* (Chichester: Wiley).

Dennett, X. and Fry, J.H.F., 1988, Overuse syndrome: a muscle biopsy study. *Lancet*, i, 905–908.

Elvey, R. L., 1979, Brachial plexus tension tests for the pathoanatomical origin of arm pain. In: *Aspects of Manipulative Therapy*, pp. 105–110 (Melbourne, Australia: Lincoln Institute of Health Sciences).

Engel, G. L., 1959, 'Psychogenic pain' and the pain-prone patient. *American Journal of Medicine*, **86**, 899–918.

Ferguson, D. A., 1971, Repetition injuries in process workers. *Medical Journal of Australia*, **2**, 408–412.

Gowers, W. R., 1888, *A Manual of Diseases of the Nervous System*, Vol. II (London: Churchill).

Grandjean, E., Nakaseko, M., Hunting, W. and Gierer, R., 1985, Researches on ergonomically designed alphanumeric keyboards. *Human Factors*, **27**(2), 175–185.

Greig, J. and Caple, D., 1986, Key activation pressure as a factor in typing techniques. *Proceedings, 23rd Annual Conference, Ergonomics Society of Australia and New Zealand*, pp. 257–265.

Hocking, B., 1989, 'Repetition strain injury' in Telecom Australia. *Medical Journal of Australia*, **150**, 724.

Hult, L., 1954, The Munkfors investigation. *Acta Orthopedica Scandinavica*, Suppl. 16: 1–76. Quoted in *Medical Management of the Regional Musculoskeletal Diseases* N. M. Hadler, 1984, (Grune and Stratton).

ISO, 1971, Alphanumeric keyboard operated with both hands. ISO 2126 (Geneva: ISO).

Malt, L. G., 1977, Keyboard design in the electronic era. Paper presented at the Printing Industry Research Association Eurotype Forum, London, September. Conference Paper No. 6.

Mathews, J., 1985, *Health and Safety at Work. Australian Trade Union Safety Representatives' Handbook* (Sydney: Pluto).

McKenzie, Robin, 1983, *Treat Your Own Neck* (Waikanae, New Zealand: Spinal Publications).

NHMRC, 1982, National Health and Medical Research Council. Approved Health Guide: repetition strain injuries. Canberra, AGPS, 1982: 1.

Patkin, M., 1984, Trying too hard. In: *Proceedings, 21st Annual Conference, Ergonomics Society of Australia and New Zealand*, pp. 298–301.

Patkin, M., 1987, Occupational strains in school children — do they exist? *Australian Educational Computing*, **2**(1), 14–22.

Patkin, M., 1988a, Excess effort and pain at work. The missing ergonomic factor. *Designing a Better World, Proceedings, 10th Congress, International Ergonomics Association, Ergonomics Society of Australia*, **1**, 248–250.

Patkin, M., 1988b, Hand and arm pain in office workers. *Modern Medicine of Australia*, **31**(10), 66–76.

Rose, M., 1986, Development of ergonomic keyboards. *Proceedings, Australian Institute of Industrial Engineering, Conference*, Sydney, 1986, pp. 17.1–17.16.

Seymour, W. G., 1966, *Industrial Skills* (London: Pitman).

Spillane, R. and Deves, L., 1987, RSI: pain, pretence, or patienthood. *Journal of Industrial Relations*, **29**, 41–48.

Stack, B., 1988a, Repetitive strain injury — prevention and rehabilitation. *Proceedings, First International Congress on Ergonomics, Occupational Health and Safety and the Environment*. Beijing, in press.

Stack, B., 1988b, New design concepts in wrist rests, desks and keyboard supports. Preprints of *International Conference on Ergonomics, Occupational Safety and Health and the Environment*, **1**, 444–453.

Taylor, R., Gow, C. and Corbett, C., 1981, *Process Workers' Arm*. Workers Health Centre, Lidcombe, New South Wales, Australia.

Welch, R., 1973, Ergonometrics of tenosynovitis. In: *Occupational Injuries. Seminar Manual*, pp. 145–147. Royal Australian College of Surgery.

Wright, G. D., 1987, The failure of the 'RSI' concept. *Medical Journal of Australia*, **147**, 233–236.

Postscript

Consequences of Ergonomic Changes in the Working Environment: Need for Systematic Analysis

THOMAS BERNS AND NIGEL CLARIDGE

Background

All measures taken to improve the working environment bring about costs in one form or another. The benefits derived from these 'costs' are not always easy to define, and are seldom considered in economic terms.

In Sweden, the majority of decision makers believe that, where necessary, improvements to the working environment should be made. However, costs which do not lead to obvious benefits, especially in economic terms, are often given low priority when allocating resources.

Ergonomics purports three main criteria:

1. safety and health,
2. comfort and well-being, and
3. productivity and efficiency.

In Sweden, traditionally, emphasis has been placed on the first two criteria, i.e. safety and health, and comfort and well-being. During the last few years, however, the need to justify improvements to the working environment from the point of view of profitability, for example by using cost–benefit analysis, has increased.

Ergonomists and work environment specialists who are actively involved with working environment issues are convinced that the majority of improvements to the work environment lead to increases in productivity. And further, as is sometimes claimed, there would appear to be no contradiction between working environment improvements and productivity. Unfortunately, there are very few studies that prove this.

In many organizations today, competition for financial resources is extremely tough. Working environment improvements have to compete with other investment policies and proposals. In order to increase the probability of success, proposed working environment investments must be prepared and presented in a similar way to other competing investment proposals. In the future, ergonomics/working environment must, to a greater extent than today, live on its own management merits and less on slogans and pressure from the outside.

However, there are substantial difficulties to overcome when carrying out a management cost–benefit analysis for working environment investments. First, there may be times when it is difficult to isolate the working environment element from other elements in a larger investment on both the cost and the benefit side. Furthermore, it may be difficult to assess the benefits when they are defined in indirect terms such as reduced absenteeism due to work-related illness, increased staff motivation and job satisfaction, as these variables are difficult to measure.

Changes in working environments affect the distribution and use of resources in different areas. Economic consequences can be defined at three different levels, depending on where and when the changes occur:

1. *Social level* — The socio-economic consequences are the consequences of working environment measures that concern society as a whole. A reduced number of work-related injuries means increased production resources for society as a whole but does not necessarily affect the profitability of the organization. Typical socio-economic consequences here are costs of medical services, occupational health services, rehabilitation and retraining programmes, and administration costs of authorities.
2. *Organization level* — The consequences with respect to industrial economics are those which affect the daily, quarterly and annual results of individual departments and the company as a whole. Usually one distinguishes between investment costs, and operating and maintenance costs.
3. *Individual level* — The effects which are here called personal economic effects are those that affect the households in connection with working environment changes or measures. These effects can have a bearing on the disposable income of the households, future development prospects, hospital charges, and so on.

Working environment changes from an industrial economic perspective

Working environments and economics — an overall view

A good working environment has for a long time been synonymous with a sufficient amount of money for environment-improving measures. When things are not running smoothly in industry, the enterprises are often accused of a lack of interest in matters relating to the working environment. Yet, results from work environment efforts need not diminish because resources are limited.

There need not be any real opposition between productivity from the point of view of industrial economics and the aim to bring about attractive working conditions and high work satisfaction. When productivity is not sufficient, monitoring and control measures are often used to increase it, usually resulting in even more intensive work dissatisfaction. This may lead to a 'negative trend' for both the company and the employees. On the other hand, good working conditions, interesting tasks and employee participation in the formulation of goals, strategy, etc., can stimulate good work achievements and even increased productivity. This may lead to a 'positive trend' for the company and the employees.

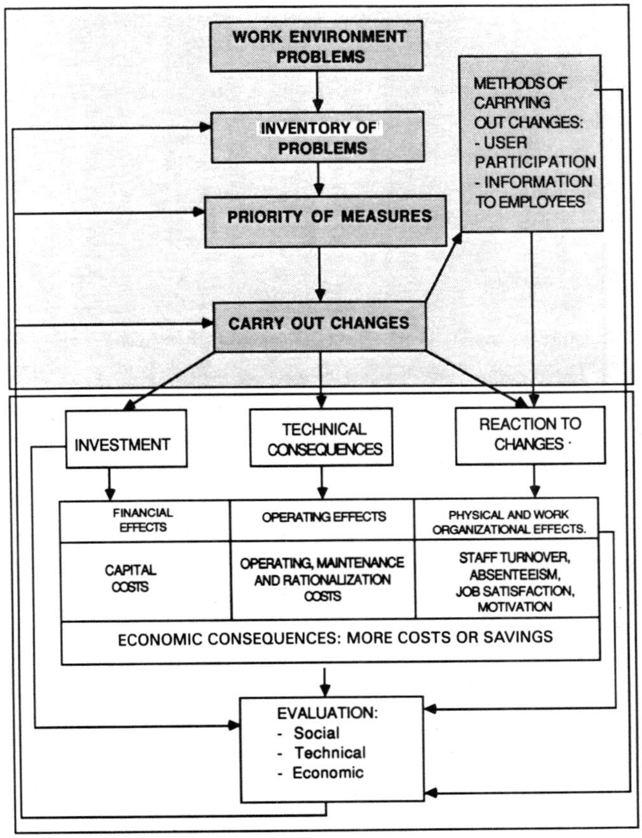

Figure 1. Economics of working environments — an overall view.

A project which aims to improve working conditions, like all other projects, runs through four main phases: (1) identifying problems, (2) listing possible measures in order of priority, (3) implementing measures, and (4) evaluating measures based on technical, economic and social data (see Figure 1).

The difficulty of evaluating working environment improvements

In practice, it is not enough to be able to describe problems and state potential returns. Management, working environment experts and personnel representatives who deal with working condition problems, must develop a decision process and a decision basis to arrange projects in order of priority. Economic arguments must be put forward based on calculation projections. Calculation often involves 'guessing' the resulting consequences, and, more importantly, knowing how to evaluate them. All cost estimates involve a degree of guesswork as they refer to the future.

It is important for all people working in the field of ergonomics, e.g. working environment experts and personnel staff, to learn to:

1. understand the language of economics in order not to fall a prey to manipulation,
2. be able to use economic perspectives — as a complement to technical, psychological and humane ones — in order to put forward important ideas, and
3. economize with limited resources.

The aim of the study

The *overall aim* of the present study was to show that good working conditions are profitable at the company level. This was achieved by collecting examples of improvements in profitability due to improved working environment conditions.

In addition, the project investigated the extent to which organizations used models for calculating working environment investments including such criteria as production volume, production quality, absenteeism due to work-related illness and staff turnover. It also investigated decision-making routines within companies based on simple cost–benefit analysis.

Planning and implementation of the work

The following work stages and data collection methods were used.

Literature studies

During the project, relevant literature was collected and used but no systematic and comprehensive literature study was conducted. Through our discussions with universities, institutes of advanced studies, research establishments, enterprises, etc., further material emerged.

Data collection by interview in organizations and administrations

The main part of the investigative work was carried out between November 1987 and March 1988. During this period we investigated the ways in which larger organizations considered the economic aspects of working environment improvements, whether the economic consequences of these improvements were calculated or estimated, which methods/models were used and which factors were included in the calculation. Some 30 large organizations were contacted. All had advanced working environment programmes and, in theory, should carry out cost–benefit analysis of planned ergonomic improvements. Organizations contacted ranged from large private enterprises, local authorities, county councils to government authorities.

Our initial contact person within the organizations was often via the occupational health service or internal working environment experts. Subsequently, contact was made with other selected persons from financial and personnel departments.

Initial contact 'sorted out' some 20 organizations. These organizations were very interested and regarded this topic as important but, surprisingly, lacked the experience, information or relevant case studies our project required. This result in itself is interesting. Interviews were conducted with key persons from the occupational health service as well as from the financial and personnel departments. Some 10 companies and organizations were represented.

The aim was to provide the project with information with respect to the economic consequences of the changes in the working environment of the selected companies and organizations. The interview survey was directed at physical as well as work organizational changes in the working environment, and also at different types of key figures and costs which were included in possible calculations.

The interviewing technique used can best be described as a 'funnel technique' with open questions. The technique means that the person interviewed is allowed to talk relatively freely about the working environment and economy and then more detailed complementary questions are asked. Each interview was structured in its entirety with the aid of a checklist drawn up in advance.

The type of information obtained in the interviews is as follows:

1. *Background data* (company, etc.).
2. *The companies' working environment programme* — It was important to obtain information about the level of ambition with regard to working environment issues within each company. Topics such as company policy, goals, strategies and action programmes were discussed.
3. *Working environment measures taken* — Companies were asked to select two projects which resulted in changes to the working environment. They were then asked whether a cost–benefit analysis had been carried out on each project.
4. *The economic consequences* — The benefits derived from an investment were discussed in terms of production, productivity, quality, rejections, staff savings, staff turnover, recruitment costs, overemployment, absenteeism due to work-related illness, training costs and capital costs.
5. *Evaluation of projects* — Project evaluation after completion can have several advantages. Learning for the future (i.e. assessing whether initial calculations were correct) is perhaps the most important, but evaluation should provide information about the effectiveness of the working environment change.
6. '*Key indices*' — Key indices selected were staff turnover, the cost per hour of long- and short-term absenteeism due to work-related illness, the cost of employing a new employee and other indices such as overtime compensation, cost of production shutdown, rejections and complaints.
7. *Effect of psychological and social factors* — Here questions were asked about the 'qualitative' aspects of working life, such as job satisfaction, motivation and so on.

Results

Our initial hypothesis was that large companies and organizations calculated or estimated the economic consequences of changes in the working environment by using methods or models which included some form of cost–benefit analysis. The hypothesis also presumed that some form of evaluation of these changes would be carried out. The results from the investigation showed that this was not the case.

Companies and organizations seldom used any systematic method of calculating or estimating the economic consequences of changes in the working environment. There was confusion about the content of and the economic value of various key indices such as absenteeism due to work-related illness and staff turnover. Further, it was unclear whether or not these indices were included in the companies' investment calculations. Also, project evaluation procedures according to economic, technical and social parameters were non-existent.

The results also showed work environment experts possessed a particularly low level of economic knowledge. A complementing factor was the low understanding of working environment issues possessed by economists, engineers and personnel management. The topics of economics/working environments seemed often to be lost in the haze between the two different subject areas.

Further, the results indicated that decision makers were not fully aware of the benefits, in both economic and ergonomic terms, to be obtained through improvements in working environment conditions. One reason for this was that costs for working environment improvements were not always easy to isolate from the entire investment costs of the programme.

Finally, the results indicated that improvements to the 'qualitative' aspects of work environment conditions, such as increased job satisfaction and motivation, could and should be estimated in economic terms. However, these aspects were judged difficult to measure and, as such, were not included in any calculations or analysis.

Conclusions

Economic investments to improve working environment conditions have not been particularly common at the social, company or individual level. In fact, no real attention has been paid to this subject until the last few years. There is now a definite trend towards an increased awareness and, in some cases, an increased need for an appraisal of this 'soft' area in economic terms.

The literature clearly indicates that many institutions, companies and individuals are becoming aware of the economic consequences of a poor working environment and of occupational injuries. Until now, the guiding principles have been of an humanitarian nature when assessing the need for preventive measures. Many people have found it offensive to evaluate working conditions and occupational injuries in economic terms. In reality, most individuals, enterprises and public authorities place some kind of value, either directly or indirectly, on people's lives and the various injuries and disabilities that are a result of a poor working environment.

The following general conclusions from the results can be made:

1. *The aim is to achieve a 'positive' trend* — Suitable economic education for working environment experts within companies and organizations is not sufficient. It is important to achieve a positive trend through good working environments,

interesting tasks at work, employee participation in the formulation of goals, and so on. This increases employee motivation which will probably lead to increased productivity.

2. *A systematic way of working is paramount* — There is a need to develop systematic working methods when evaluating the economic benefits of working environment investments. Where the aim has been to analyse the economic aspects of working environment investments, calculations and decisions have been based upon 'qualified feelings' rather than by well-prepared, systematic approaches. There is a considerable need for a systematic working method that integrates working environment issues in the daily economic and budget routines within companies and organizations.
3. *Working environment experts who 'think economics' are worth their weight in gold* — In an enterprise there are many different professional languages but the common language is economics. It is important that working environment experts within companies understand and can argue for different working environment measures in economic terms. In order to do this, it is necessary to understand industrial economics and also to be able to relate proposals for changes in the working environment to the economy of the company. Economic criteria should not substitute for other criteria, such as health, safety and well-being, but should complement the decision basis.
4. *Managers play a key part in work environmental work* — Working environment issues must become an integral part of general management. The efficiency of routines and working methods in the field of working environments, likewise the quality of the final working environment, can and will never be better than the manager's ability and level of ambition within the area.

   If proposals for improving the working environment are to become accepted in practice, work environment issues must be integrated in the budget and decision process of companies more efficiently than exists today. Otherwise, changes to the working environment will always be regarded as a cost and a necessary evil, and not as part of a modern organization.
5. *Economic follow-up of changes in the working environment is important* — The need to follow up and evaluate improvements to the work environment is great. Evaluation criteria can be defined in economic, technical and social terms as laid down in the original project specification. This must take place in order to obtain an experience for future proposals and calculations, and to be able to follow up implemented measures from an economic viewpoint.
6. *The concept of 'key indices' must be clarified* — Clarifying what is meant by 'key indices' and how they are used in the daily routines of companies are the two major areas requiring attention. Those factors which companies include when defining staff turnover, absenteeism due to work-related illness, and so on, can vary significantly. For example, the estimated cost for recruiting a shopfloor employee can vary from 30 000 to 90 000 Swedish crowns within different companies. This makes calculation work somewhat precarious. There is a need to clarify the concept of 'key indices' and establish values which are generally acceptable or which are specific to certain types of industry.

Bibliography

Ahlmann, H., 1987, Productive work organisations. In: *Människan i Arbete*, pp. 584–588.

Andreoni, D., 1986, *The Cost of Occupational Accidents and Diseases*. Occupational Safety and Health Series, No. 54 (Geneva: International Labour Office).

The Employers Association, 1987, Humane environment — a profitable investment. SFO 215 8701.

Bergmark, U., 1988, How much does an employee cost? *Administration*, **4**.

Dahlberg, Å. and Engstrand, G., 1982, Social aspects of investments in the work environment. *Nordiska Ergonomisällskapets konferens — Ergonomi och lönsamhet*.

Eckerhall, A., 1988, Work injury insurance. Reforms can save millions. *Dagens Industri*, 23 February.

Frick, K., 1986, Economic evaluation as a support for local work environment routines. *Arbete-Människa-Miljö*, **1**, 3–10.

Gröjer, J.-E. and Johansson, U., *Result Orientated PA. Economics for Personnel Issues* (Stockholm: Liber).

Herring, V. and Berns, T., 1983, *Cost/Benefits of Ergonomics. A Literature Review*. ERGOLAB Internal Report.

Högström, M. and Nilsen, P., 1988, *The Safety Engineer — A Profitable Consultant* (Stockholm: SAF).

Johansson, U., 1987, *To Develop the Human Capital* (Stockholm: Svenska Personalföreningen) (SPF).

Liukkonen, P., 1987, An economic evaluation of work environment improvements to a forge. Stockholm University.

Index